Biophysics:
An Introduction

Biophysics:
An Introduction

Edited by
Zane Bradley

目 Larsen & Keller
www.larsen-keller.com

Biophysics: An Introduction
Edited by Zane Bradley
ISBN: 978-1-63549-048-0 (Hardback)

☰ Larsen & Keller

Published by Larsen and Keller Education,
5 Penn Plaza,
19th Floor,
New York, NY 10001, USA

Cataloging-in-Publication Data

Biophysics : an introduction / edited by Zane Bradley.
    p. cm.
Includes bibliographical references and index.
ISBN 978-1-63549-048-0
1. Biophysics. 2. Biology. 3. Physics. I. Bradley, Zane.
QH505 .B56 2017
571.4--dc23

The publisher's policy is to use permanent paper from mills that operate a sustainable forestry policy. Furthermore, the publisher ensures that the text paper and cover boards used have met acceptable environmental accreditation standards.

Printed and bound in the United States of America.

For more information regarding Larsen and Keller Education and its products, please visit the publisher's website www.larsen-keller.com

Table of Contents

Preface

Biophysics is that branch of sciences which deals with the study of biological processes using physical laws, theories and methods. This field overlaps with nanotechnology, systems biology, biochemistry, bioengineering and computational biology. This book outlines the processes and applications of this subject in detail. It is a valuable compilation of topics, ranging from the basic to the most complex theories and principles in the field of biophysics. Coherent flow of topics, student-friendly language and extensive use of examples make this text an invaluable source of knowledge. The textbook aims to shed light on some of the unexplored concepts of biophysics.

A foreword of all Chapters of the book is provided below:

Chapter 1 - Biophysics studies the approaches and methods that are used to study biological systems. It covers the majority of biological organizations, these organizations vary from molecular to organismic. The chapter on biophysics offers an insightful focus, keeping in mind the subject matter; **Chapter 2** - The methods and techniques are an important component of any field of study. Some of these methods and techniques are biophotonics, calcium imaging, patch clamp, calorimetry, chromatography, microscopy and electron microscope. The aspects elucidated in this section are of vital importance and provides a better understanding of biophysics; **Chapter 3** - The allied fields of biophysics are medical biophysics, clinical biophysics, membrane biophysics, molecular biophysics and biophysical chemistry. Medical physics is the application of physics and the methods that are used in medicine. This chapter is a compilation of the various branches of biophysics that form an integral part of the subject matter; **Chapter 4** - The study of chemical reactions is known as enzyme kinetics. Enzymes activate these reactions and act as catalysts. Molecular motor, myoglobin, hemoglobin, pancreatic ribonuclease and antibody are some of the features explained in the following section. The section strategically encompasses and incorporates the main aspects of biophysical structure and phenomena, providing a complete understanding; **Chapter 5** - Quantum biology is the application of quantum mechanics to objects and problems. The processes involved are chemical reactions, formation of excited electronic states and transfer of excitation energy. The topics explained in the following chapter are visual phototransduction, magnetoreception, orchestrated objective reduction and biomagnetism; **Chapter 6** - The related aspects of biophysics are known as lipid polymorphism, biophysical profile, medical physics and cell biophysics. Lipid polymorphism is the capability of lipids to aggregate in different ways to form structures. The topics discussed in the chapter are of great importance to broaden the existing knowledge on biophysics.

I would like to thank the entire editorial team who made sincere efforts for this book and my family who supported me in my efforts of working on this book. I take this opportunity to thank all those who have been a guiding force throughout my life.

Editor

Introduction to Biophysics

Biophysics studies the approaches and methods that are used to study biological systems. It covers the majority of biological organizations, these organizations vary from molecular to organismic. The chapter on biophysics offers an insightful focus, keeping in mind the subject matter.

Biophysics is an interdisciplinary science that applies the approaches and methods of physics to study biological systems. Biophysics covers all scales of biological organization, from molecular to organismic and populations. Biophysical research shares significant overlap with biochemistry, nanotechnology, bioengineering, computational biology, biomechanics and systems biology.

The term *biophysics* was originally introduced by Karl Pearson in 1892.

Overview

Molecular biophysics typically addresses biological questions similar to those in biochemistry and molecular biology, but more quantitatively, seeking to find the physical underpinnings of biomolecular phenomena. Scientists in this field conduct research concerned with understanding the interactions between the various systems of a cell, including the interactions between DNA, RNA and protein biosynthesis, as well as how these interactions are regulated. A great variety of techniques are used to answer these questions.

Fluorescent imaging techniques, as well as electron microscopy, x-ray crystallography, NMR spectroscopy, atomic force microscopy (AFM) and small-angle scattering (SAS) both with X-rays and neutrons (SAXS/SANS) are often used to visualize structures of biological significance. Protein dynamics can be observed by neutron spin echo spectroscopy. Conformational change in structure can be measured using techniques such as dual polarisation interferometry, circular dichroism, SAXS and SANS. Direct manipulation of molecules using optical tweezers or AFM, can also be used to monitor biological events where forces and distances are at the nanoscale. Molecular biophysicists often consider complex biological events as systems of interacting entities which can be understood e.g. through statistical mechanics, thermodynamics and chemical kinetics. By drawing knowledge and experimental techniques from a wide variety of disciplines, biophysicists are often able to directly observe, model or even manipulate the structures and interactions of individual molecules or complexes of molecules.

In addition to traditional (i.e. molecular and cellular) biophysical topics like structural biology or enzyme kinetics, modern biophysics encompasses an extraordinarily broad range of research, from bioelectronics to quantum biology involving both experimental and theoretical tools. It is becoming increasingly common for biophysicists to apply the models and experimental techniques derived from physics, as well as mathematics and statistics, to larger systems such as tissues, organs, populations and ecosystems. Biophysical models are used extensively in the study of electrical conduction in single neurons, as well as neural circuit analysis in both tissue and whole brain.

History

Some of the earlier studies in biophysics were conducted in the 1840s by a group known as the Berlin school of physiologists. Among its members were pioneers such as Hermann von Helmholtz, Ernst Heinrich Weber, Carl F. W. Ludwig, and Johannes Peter Müller. Biophysics might even be seen as dating back to the studies of Luigi Galvani.

The popularity of the field rose when the book *What Is Life?* by Erwin Schrödinger was published. Since 1957 biophysicists have organized themselves into the Biophysical Society which now has about 9,000 members over the world.

Focus as a Subfield

While some colleges and universities have dedicated departments of biophysics, usually at the graduate level, many do not have university-level biophysics departments, instead having groups in related departments such as molecular biology, biochemistry, chemistry, computer science, mathematics, medicine, pharmacology, physiology, physics, and neuroscience. Depending on the strengths of a department at a university differing emphasis will be given to fields of biophysics. What follows is a list of examples of how each department applies its efforts toward the study of biophysics. This list is hardly all inclusive. Nor does each subject of study belong exclusively to any particular department. Each academic institution makes its own rules and there is much overlap between departments.

- Biology and molecular biology – Almost all forms of biophysics efforts are included in some biology department somewhere. Typical examples include: gene regulation, single protein dynamics, bioenergetics, patch clamping, biomechanics, virophysics.

- Structural biology – Ångstrom-resolution structures of proteins, nucleic acids, lipids, carbohydrates, and complexes thereof.

- Biochemistry and chemistry – biomolecular structure, siRNA, nucleic acid structure, structure-activity relationships.

- Computer science – Neural networks, biomolecular and drug databases.

- Computational chemistry – molecular dynamics simulation, molecular docking, quantum chemistry

- Bioinformatics – sequence alignment, structural alignment, protein structure prediction

- Mathematics – graph/network theory, population modeling, dynamical systems, phylogenetics.

- Medicine, neuroscience and neurophysics – studying neural networks experimentally (brain slicing) as well as theoretically (computer models), membrane permitivity, gene therapy, understanding tumors.

- Pharmacology and physiology – channelomics, biomolecular interactions, cellular membranes, polyketides.

- Physics – negentropy, stochastic processes, covering dynamics.

- Quantum biology – The field of quantum biology applies quantum mechanics to biological objects and problems

decoheredisomers to yield time-dependent base substitutions. These studies imply applications in quantum computing.

- Agronomy and agriculture

Many biophysical techniques are unique to this field. Research efforts in biophysics are often initiated by scientists who were traditional physicists, chemists, and biologists by training.

References

- Donald R. Franceschetti. Applied Science - 5 Volume Set. SALEM PressINC; 15 May 2012. ISBN 978-1-58765-781-8. p. 234.

- Joe Rosen; Lisa Quinn Gothard. Encyclopedia of Physical Science. Infobase Publishing; 2009. ISBN 978-0-8160-7011-4. p. 49.

- Hobbie RK, Roth BJ (2006). Intermediate Physics for Medicine and Biology (4th ed.). Springer. ISBN 978-0-387-30942-2.

Biophysics: Methods and Techniques

2

The methods and techniques are an important component of any field of study. Some of these methods and techniques are biophotonics, calcium imaging, patch clamp, calorimetry, chromatography, microscopy and electron microscope. The aspects elucidated in this section are of vital importance and provides a better understanding of biophysics.

Biophotonics

The term biophotonics denotes a combination of biology and photonics, with photonics being the science and technology of generation, manipulation, and detection of photons, quantum units of light. Photonics is related to electronics and photons. Photons play a central role in information technologies such as fiber optics the way electrons do in electronics.

Biophotonics can also be described as the "development and application of optical techniques, particularly imaging, to the study of biological molecules, cells and tissue". One of the main benefits of using optical techniques which make up biophotonics is that they preserve the integrity of the biological cells being examined.

Biophotonics has therefore become the established general term for all techniques that deal with the interaction between biological items and photons. This refers to emission, detection, absorption, reflection, modification, and creation of radiation from biomolecular, cells, tissues, organisms and biomaterials. Areas of application are life science, medicine, agriculture, and environmental science. Similar to the differentiation between "electric" and "electronics" a difference can be made between applications such as Therapy and surgery, which use light mainly to transfer energy, and applications such as diagnostics, which use light to excite matter and to transfer information back to the operator. In most cases the term biophotonics refers to the latter type of application.

Applications

Biophotonics can be used to study biological materials or materials with properties similar to biological material, i.e., scattering material, on a microscopic or macroscopic scale. On the microscopic scale common applications include microscopy and optical coherence tomography. On the macroscopic scale, the light is diffuse and applications commonly deal with diffuse optical imaging and tomography (DOI and DOT).

In microscopy, the development and refinement of the confocal microscope, the fluorescence microscope, and the total internal reflection fluorescence microscope all belong to the field of biophotonics.

The specimens that are imaged with microscopic techniques can also be manipulated by optical tweezers and laser micro-scalpels, which are further applications in the field of biophotonics.

DOT is a method used to reconstruct an internal anomaly inside a scattering material. The method is noninvasive and only requires the data collected at the boundaries. The typical procedure involves scanning a sample with a light source while collecting light that exits the boundaries. The collected light is then matched with a model, for example, the diffusion model, giving an optimization problem.

FRET

Fluorescence Resonance Energy Transfer, also known as Foerster Resonance Energy Transfer (FRET in both cases) is the term given to the process where two excited "fluorophores" pass energy one to the other non-radiatively (i.e., without exchanging a photon). By carefully selecting the excitation of these flurophores and detecting the emission, FRET has become one of the most widely used techniques in the field of Biophotonics, giving scientists the chance to investigate sub-cellular environments.

Light Sources

The predominantly used light sources are beam lights. LEDs and superluminescent diodes also play an important role. Typical wavelengths used in biophotonics are between 600 nm (Visible) and 3000 nm (near IR).

Lasers

Lasers play an increasingly important role in biophotonics. Their unique intrinsic properties like precise wavelength selection, widest wavelength coverage, highest focusability and thus best spectral resolution, strong power densities and broad spectrum of excitation periods make them the most universal light tool for a wide spectrum of applications. As a consequence a variety of different laser technologies from a broad number of suppliers can be found in the market today.

Gas Lasers

Major gas lasers used for biophotonics applications, and their most important wavelengths, are:

- Argon Ion laser: 457.8 nm, 476.5 nm, 488.0 nm, 496.5 nm, 501.7 nm, 514.5 nm (multi-·tion possible)

- Krypton Ion laser: 350.7 nm, 356.4 nm, 476.2 nm, 482.5 nm, 520.6 nm, 530.9 nm, 568.2 nm, 647.1 nm, 676.4 nm, 752.5 nm, 799.3 nm

- Helium–neon laser: 632.8 nm (543.5 nm, 594.1 nm, 611.9 nm)

- HeCd lasers: 325 nm, 442 nm

Other commercial gas lasers like carbon dioxide (CO_2), carbon monoxide, nitrogen, oxygen, xenon-ion, excimer or metal vapor lasers have no or only very minor importance in biophotonics. Major advantage of gas lasers in biophotonics is their fixed wavelength, their perfect beam quality and their low linewidth/high coherence. Argon ion lasers can also operate in multi-line mode. Major disadvantage are high power consumption, generation of mechanical noise due to fan cooling and limited laser powers. Key suppliers are Coherent, CVI/Melles Griot, JDSU, Lasos, LTB and Newport/Spectra Physics.

Diode Lasers

The most commonly integrated laser diodes, which are used for diode lasers in biophotonics are based either on GaN or GaAs semiconductor material. GaN covers a wavelength spectrum from 375 to 488 nm (commercial products at 515 have been announced recently) whereas GaAs covers a wavelength spectrum starting from 635 nm.

Most commonly used wavelengths from diode lasers in biophotonics are: 375, 405, 445, 473, 488, 515, 640, 643, 660, 675, 785 nm.

Laser Diodes are available in 4 classes:

- Single edge emitter/broad stripe/broad area

- Surface emitter/VCSEL

- Edge emitter/Ridge waveguide

- Grating stabilized (FDB, DBR, ECDL)

For biophotonic applications, the most commonly used laser diodes are edge emitting/ridge waveguide diodes, which are single transverse mode and can be optimized to an almost perfect TEM00 beam quality. Due to the small size of the resonator, digital modulation can be very fast (up to 500 MHz). Coherence length is low (typically < 1 mm) and the typical linewidth is in the nm-range. Typical power levels are around 100 mW (depending on wavelength and supplier). Key suppliers are: Coherent, Melles Griot, Omicron, Toptica, JDSU, Newport, Oxxius, Power Technology. Grating stabilized diode lasers either have an lithographical incorporated grating (DFB, DBR) or an external grating (ECDL). As a result, the coherence length will raise into the range of several meters, whereas the linewidth will drop well below picometers (pm). Biophotonic applications, which make use of this characteristics are Raman spectroscopy (requires linewidth below cm-1) and spectroscopic gas sensing.

Solid State Lasers

Solid-state lasers are lasers based on solid-state gain media such as crystals ᴖ
es doped with rare earth or transition metal ions, or semiconductor lasers. (Altᴗ
semiconductor lasers are of course also solid-state devices, they are often not inclᴗ
ed in the term solid-state lasers.) Ion-doped solid-state lasers (also sometimes calleᴅ
doped insulator lasers) can be made in the form of bulk lasers, fiber lasers, or other
types of waveguide lasers. Solid-state lasers may generate output powers between a few
milliwatts and (in high-power versions) many kilowatts.

Ultrachrome Lasers

Many advanced applications in biophotonics require individually selectable light at
multiple wavelengths. As a consequence a series of new laser technologies has been
introduced, which currently looks for precise wording.

The most commonly used terminology are supercontinuum lasers, which emit visible
light over a wide spectrum simultaneously. This light is then filtered e.g. via acousto-optic
modulators (AOM, AOTF) into 1 or up to 8 different wavelengths. Typical suppliers for
this technology was NKT Photonics or Fianium. Recently NKT Photonics bought Fianium, remaining the major supplier of the supercontinuum technology on the market.

In another approach (Toptica/iChrome) the supercontinuum is generated in the in-
fra-red and then converted at a single selectable wavelength into the visible regime.
This approach does not require AOTF's and has a background-free spectral purity.

Since both concepts have major importance for biophotonics the umbrella term "ultra-
chrome lasers" is often used.

OPOs

Single Photon Sources

Single photon sources are novel types of light sources distinct from coherent light
sources (lasers) and thermal light sources (such as incandescent light bulbs and mercury-vapor lamps) that emit light as single particles or photons.

Calcium Imaging

Calcium imaging is a scientific technique usually carried out in research which is designed to show the calcium (Ca^{2+}) status of a cell, tissue or medium. Calcium imaging
techniques take advantage of so-called calcium indicators, fluorescent molecules that
can respond to the binding of Ca^{2+} ions by changing their fluorescence properties. Two

.nain classes of calcium indicators exist: chemical indicators and genetically encoded calcium indicators (GECI). Calcium imaging can be used to optically probe intracellular calcium in living animals. This technique has allowed studies of Ca^{2+} signalling in a wide variety of cell types and neuronal activity in hundreds of neurons and glial cells within neuronal circuits .

Chemical Indicators

Chemical indicators are small molecules that can chelate calcium ions. All these molecules are based on an EGTA homologue called BAPTA, with high selectivity for calcium (Ca^{2+}) ions versus magnesium (Mg^{2+}) ions.

This group of indicators includes fura-2, indo-1, fluo-3, fluo-4, Calcium Green-1.

These dyes are often used with the chelator carboxyl groups masked as acetoxymethyl esters, in order to render the molecule lipophilic and to allow easy entrance into the cell. Once this form of the indicator is in the cell, cellular esterases will free the carboxyl groups and the indicator will be able to bind calcium. The free acid form of the dyes (i.e. without the acetoxymethyl ester modification) can also be directly injected into cells via a microelectrode or micropipette which removes uncertainties as to the cellular compartment holding the dye (the acetoxymethyl ester can also enter the endoplasmic reticulum and mitochondria). Binding of a Ca^{2+} ion to a fluorescent indicator molecule leads to either an increase in quantum yield of fluorescence or emission/excitation wavelength shift. Individual chemical Ca^{2+} fluorescent indicators are utilized for cytosolic calcium measurements in a wide variety of cellular preparations. The first real time (video rate) Ca^{2+} imaging was carried out in 1986 in cardiac cells using intensified video cameras. Later development of the technique using laser scanning confocal microscopes revealed sub-cellular Ca^{2+} signals in the form of Ca^{2+} sparks and Ca^{2+} blips. Relative responses from a combination of chemical Ca^{2+} fluorescent indicators were also used to quantify calcium transients in intracellular organelles such as mitochondria.

Calcium imaging, also referred to as calcium mapping, is also used to perform research on myocardial tissue. Calcium mapping is a ubiqutous technique used on whole, isolated hearts such as mouse, rat, and rabbit species.

Genetically Encoded Calcium Indicator

These indicators are fluorescent proteins derived from green fluorescent protein (GFP) or its variants (e.g. circularly permuted GFP, YFP, CFP), fused with calmodulin (CaM) and the M13 domain of the myosin light chain kinase, which is able to bind CaM. Alternatively, variants of GFP are fused with the calcium binding protein troponin C (TnC), applying the mechanism of FRET (Förster Resonance Energy Transfer) for signal modulation.

Genetically encoded indicators do not need to be loaded onto cells, instead the genes encoding for these proteins can be easily transfected to cell lines. It is also possible to create transgenic animals expressing the dye in all cells or selectively in certain cellular subtypes.

Examples of genetically encoded indicators are Pericams, Cameleons, GCaMP, TN-XXL and Twitch.

Usage

Regardless of the type of indicator used the imaging procedure is generally very similar. Cells loaded with an indicator or expressing it in the case of GECI, can be viewed using a fluorescence microscope and captured by a CCD camera. Confocal and two-photon microscopes provide improved sectioning ability so that calcium signals can be resolved in micro domains or (for example) synaptic boutons. Images are analyzed by measuring fluorescence intensity changes for a single wavelength or two wavelengths expressed as a ratio (ratiometric indicators). The derived fluorescence intensities and ratios are plotted against calibrated values for known Ca^{2+} levels to learn Ca^{2+} concentration.

Patch Clamp

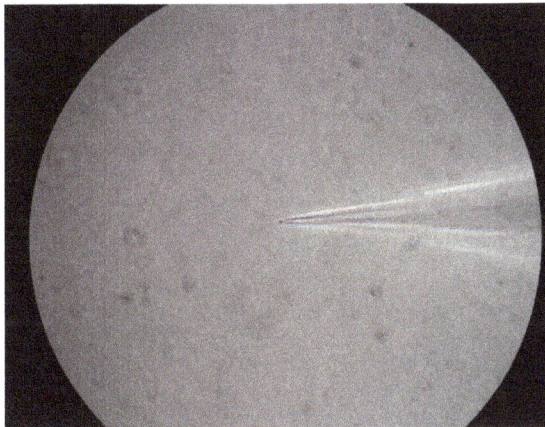

A bacterial spheroplast patched with a glass pipette

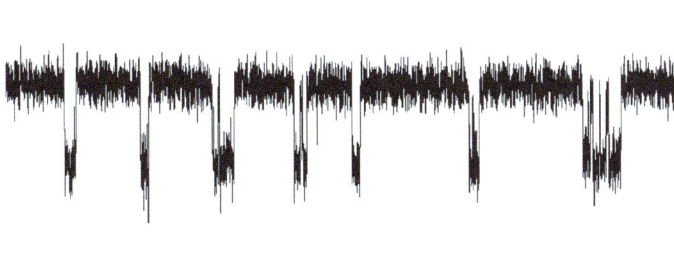

A patch clamp recording of current reveals transitions between two conductance states of a single ion channel: closed (at top) and open (at bottom).

The patch clamp technique is a laboratory technique in electrophysiology that allows the study of single or multiple ion channels in cells. The technique can be applied to a wide variety of cells, but is especially useful in the study of excitable cells such as neurons, cardiomyocytes, muscle fibers, and pancreaticbeta cells. It can also be applied to the study of bacterial ion channels in specially prepared giant spheroplasts.

The patch clamp technique is a refinement of the voltage clamp. Erwin Neher and Bert Sakmann developed the patch clamp in the late 1970s and early 1980s. This discovery made it possible to record the currents of single ion channel molecules for the first time, which improved understanding of the involvement of channels in fundamental cell processes such as action potentials and nerve activity. Neher and Sakmann received the Nobel Prize in Physiology or Medicine in 1991 for this work.

Basic Technique

Set-up

Classical patch clamp setup, with microscope, antivibration table, and micromanipulators

Patch clamp recording uses a glass micropipette called a patch pipette as a recording electrode, and another electrode in the bath around the cell, as a reference ground electrode. Depending on what the researcher is trying to measure, the diameter of the pipette tip used may vary, but it is usually in the micrometer range. This small size is used to enclose a membrane surface area or "patch" that often contains just one or a few ion channel molecules. This type of electrode is distinct from the "sharp microelectrode" used to puncture cells in traditional intracellular recordings, in that it is sealed onto the surface of the cell membrane, rather than inserted through it.

Micropipette Pulling Device
(Vertical Type)

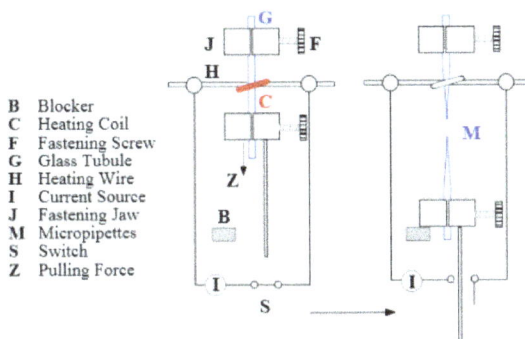

B   Blocker
C   Heating Coil
F   Fastening Screw
G   Glass Tubule
H   Heating Wire
I   Current Source
J   Fastening Jaw
M   Micropipettes
S   Switch
Z   Pulling Force

Sequence: fasten
                 heat
                 pull / separate

Schematic depiction of a pipette puller device used to
prepare micropipettes for patch clamp and other recordings

In some experiments, the micropipette tip is heated in a microforge to produce a smooth surface that assists in forming a high resistance seal with the cell membrane. To obtain this high resistance seal, the micropipette is pressed against a cell membrane and suction is applied. A portion of the cell membrane is suctioned into the pipette, creating an omega-shaped area of membrane which, if formed properly, creates a resistance in the 10–100 gigaohms range, called a "gigaohm seal" or "gigaseal". The high resistance of this seal makes it possible to isolate electronically the currents measured across the membrane patch with little competing noise, as well as providing some mechanical stability to the recording.

Depending on the experiment, the interior of the pipette can be filled with a solution matching the ionic composition of the bath solution, as in the case of cell-attached recording, or matching the cytoplasm, for whole-cell recording. The researcher can also change the content or concentration of these solutions by adding ions or drugs to study the ion channels under different conditions.

Recording

Many patch clamp amplifiers do not use true voltage clamp circuitry, but instead are differential amplifiers that use the bath electrode to set the zero current (ground) level. This allows a researcher to keep the voltage constant while observing changes in current. To make these recordings, the patch pipette is compared to the ground electrode. Current is then injected into the system to maintain a constant, set voltage. However much current is needed to clamp the voltage is opposite in sign and equal in magnitude to the current through the membrane.

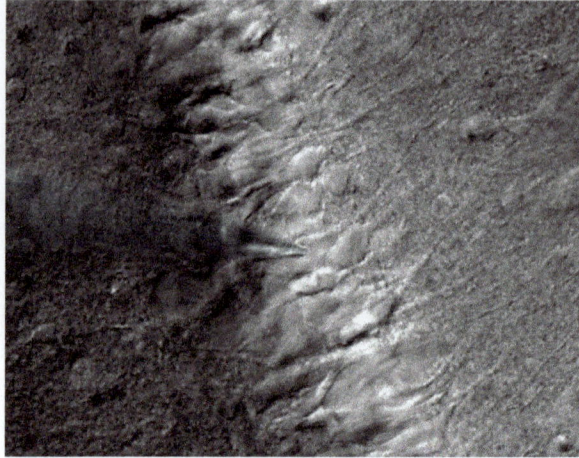

Patch clamp of a nerve cell within a slice of brain tissue.
The pipette in the photograph has been marked with a slight blue color.

Alternatively, the cell can be current clamped in whole-cell mode, keeping current constant while observing changes in membrane voltage.

Variations

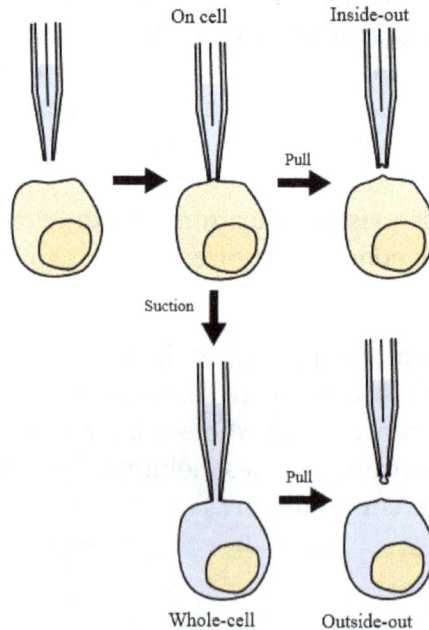

Diagram showing variations of the patch clamp technique

Several variations of the basic technique can be applied, depending on what the researcher wants to study. The inside-out and outside-out techniques are called "excised patch" techniques, because the patch is excised (removed) from the main body of the cell. Cell-attached and both excised patch techniques are used to study the behavior of individual ion channels in the section of membrane attached to the electrode.

Whole-cell patch and perforated patch allow the researcher to study the electrical behavior of the entire cell, instead of single channel currents. The whole-cell patch, which enables low-resistance electrical access to the inside of a cell, has now largely replaced high-resistance microelectrode recording techniques to record currents across the entire cell membrane.

Cell-attached Patch

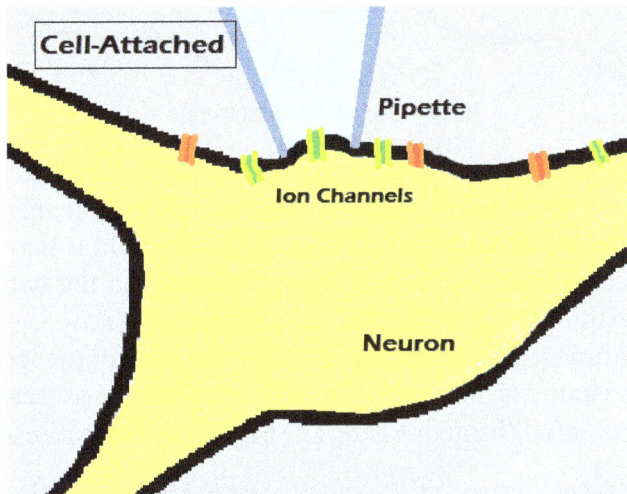

Cell-attached patch configuration

For this method, the pipette is sealed onto the cell membrane to obtain a gigaseal, while ensuring that the cell membrane remains intact. This allows the recording of currents through single, or a few, ion channels contained in the patch of membrane captured by the pipette. By only attaching to the exterior of the cell membrane, there is very little disturbance of the cell structure. Also, by not disrupting the interior of the cell, any intracellular mechanisms normally influencing the channel will still be able to function as they would physiologically. Using this method it is also relatively easy to obtain the right configuration, and once obtained it is fairly stable.

For ligand-gated ion channels or channels that are modulated by metabotropic receptors, the neurotransmitter or drug being studied is usually included in the pipette solution, where it can interact with what used to be the external surface of the membrane. The resulting channel activity can be attributed to the drug being used, although it is usually not possible to then change the drug concentration inside the pipette. The technique is thus limited to one point in a dose response curve per patch. Therefore, the dose response is accomplished using several cells and patches. However, voltage-gated ion channels can be clamped successively at different membrane potentials in a single patch. This results in channel activation as a function of voltage, and a complete I-V (current-voltage) curve can be established in only one patch. Another potential drawback of this technique is that, just as the intracellular pathways of the cell are not disturbed, they cannot be directly modified either.

Inside-out Patch

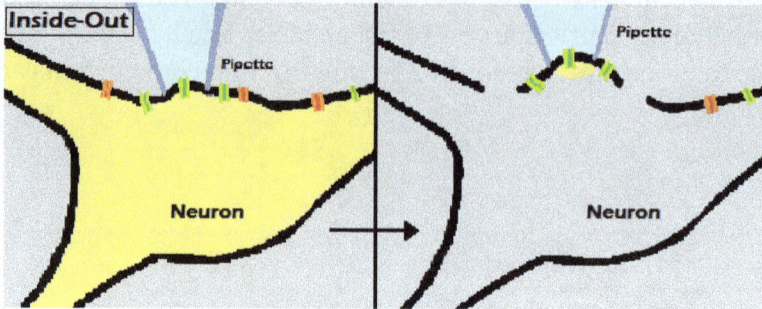

Inside-out patch configuration

In the inside-out method, a patch of the membrane is attached to the patch pipette, detached from the rest of the cell, and the cytosolic surface of the membrane is exposed to the external media, or bath. One advantage of this method is that the experimenter has access to the intracellular surface of the membrane via the bath and can change the chemical composition of what the surface of the membrane is exposed to. This is useful when an experimenter wishes to manipulate the environment at the intracellular surface of single ion channels. For example, channels that are activated by intracellular ligands can then be studied through a range of ligand concentrations.

To achieve the inside-out configuration, the pipette is attached to the cell membrane as in the cell-attached mode, forming a gigaseal, and is then retracted to break off a patch of membrane from the rest of the cell. Pulling off a membrane patch often results initially in the formation of a vesicle of membrane in the pipette tip, because the ends of the patch membrane fuse together quickly after excision. The outer face of the vesicle must then be broken open to enter into inside-out mode; this may be done by briefly taking the membrane through the bath solution/air interface, by exposure to a low Ca^{2+} solution, or by momentarily making contact with a droplet of paraffin or a piece of cured silicone polymer.

Whole-cell Recording or Whole-cell Patch

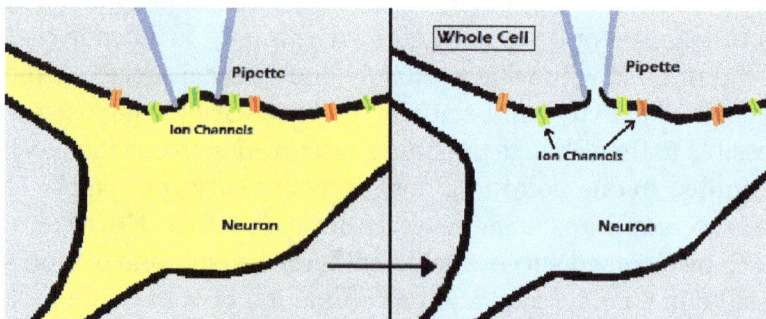

Whole-cell patch configuration

Whole-cell recordings involve recording currents through multiple channels simultaneously, over the membrane of the entire cell. The electrode is left in place on the cell, as in cell-attached recordings, but more suction is applied to rupture the membrane patch, thus providing access from the interior of the pipette to the intracellular space of the cell. Once the pipette is attached to the cell membrane, there are two methods of breaking the patch. The first is by applying more suction. The amount and duration of this suction depends on the type of cell and size of the pipette. The other method requires a large current pulse to be sent through the pipette. How much current is applied and the duration of the pulse also depend on the type of cell. For some types of cells, it is convenient to apply both methods simultaneously to break the patch.

The advantage of whole-cell patch clamp recording over sharp microelectrode recording is that the larger opening at the tip of the patch clamp electrode provides lower resistance and thus better electrical access to the inside of the cell. A disadvantage of this technique is that because the volume of the electrode is larger than the volume of the cell, the soluble contents of the cell's interior will slowly be replaced by the contents of the electrode. This is referred to as the electrode "dialyzing" the cell's contents. After a while, any properties of the cell that depend on soluble intracellular contents will be altered. The pipette solution used usually approximates the high-potassium environment of the interior of the cell to minimize any changes this may cause. There is often a period at the beginning of a whole-cell recording when one can take measurements before the cell has been dialyzed.

Outside-out Patch

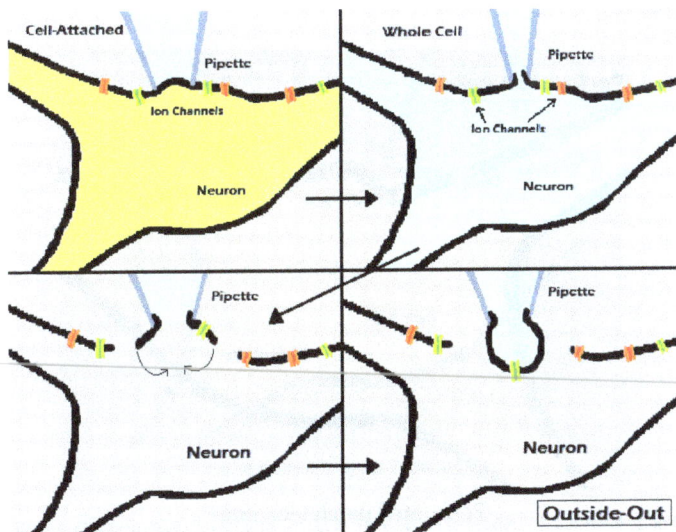

Outside-out patch formation technique. In order: top-left, top-right, bottom-left, bottom-right

The name "outside-out" emphasizes both this technique's complementarity to the inside-out technique, and the fact that it places the external rather than intracellular surface of the cell membrane on the outside of the patch of membrane, in relation to the patch electrode.

To achieve a loose patch clamp on a cell membrane, the pipette is moved slowly towards the cell, until the electrical resistance of the contact between the cell and the pipette increases to a few times greater resistance than that of the electrode alone. The closer the pipette gets to the membrane, the greater the resistance of the pipette tip becomes, but if too close a seal is formed, and it could become difficult to remove the pipette without damaging the cell. For the loose patch technique, the pipette does not get close enough to the membrane to form a gigaseal or a permanent connection, nor to pierce the cell membrane. The cell membrane stays intact, and the lack of a tight seal creates a small gap through which ions can pass outside the cell without entering the pipette.

A significant advantage of the loose seal is that the pipette that is used can be repeatedly removed from the membrane after recording, and the membrane will remain intact. This allows repeated measurements in a variety of locations on the same cell without destroying the integrity of the membrane. This flexibility has been especially useful to researchers for studying muscle cells as they contract under real physiological conditions, obtaining recordings quickly, and doing so without resorting to drastic measures to stop the muscle fibers from contracting. A major disadvantage is that the resistance between the pipette and the membrane is greatly reduced, allowing current to leak through the seal, and significantly reducing the resolution of small currents. This leakage can be partially corrected for, however, which offers the opportunity to compare and contrast recordings made from different areas on the cell of interest. Given this, it has been estimated that the loose patch technique can resolve currents smaller than 1 mA/cm^2.

Automatic Patch Clamping

Automated patch clamp systems have recently been developed, in order to collect large amounts of data inexpensively in a shorter period of time. Such systems typically include a single-use microfluidic device, either an injection molded or a polydimethylsiloxane (PDMS) cast chip, to capture a cell or cells, and an integrated electrode.

In one form of such an automated system, a pressure differential is used to force the cells being studied to be drawn towards the pipette opening until they form a gigaseal. Then, by briefly exposing the pipette tip to the atmosphere, the portion of the membrane protruding from the pipette bursts, and the membrane is now in the inside-out conformation, at the tip of the pipette. In a completely automated system, the pipette and the membrane patch can then be rapidly moved through a series of different test solutions, allowing different test compounds to be applied to the intracellular side of the membrane during recording.

Neutron Spin Echo

Neutron spin echo spectroscopy is an inelastic neutron scattering technique invented by Ferenc Mezei in the 1970s, and developed in collaboration with John Hayter. In recognition of his work and in other areas, Mezei was awarded the first Walter Haelg Prize in 1999.

The spin echo spectrometer possesses an extremely high energy resolution (roughly one part in 100,000). Additionally, it measures the density-density correlation (or intermediate scattering function) F(Q,t) as a function of momentum transfer Q and time. Other neutron scattering techniques measure the dynamic structure factor S(Q,ω), which can be converted to F(Q,t) by a Fourier transform, which may be difficult in practice. For weak inelastic features S(Q,ω) is better suited, however, for (slow) relaxations the natural representation is given by F(Q,t). Because of its extraordinary high effective energy resolution compared to other neutron scattering techniques, NSE is an ideal method to observe overdamped internal dynamic modes (relaxations) and other diffusive processes in materials such as a polymer blends, alkane chains, or microemulsions. The extraordinary power of NSE spectrometry was further demonstrated recently by the direct observation of coupled internal protein dynamics in the proteinsNHERF1 and Taq polymerase, allowing the direct visualization of protein nanomachinery in motion.

How it Works

Neutron spin echo is a time-of-flight technique. Concerning the neutron spins it has a strong analogy to the so-called Hahn echo, well known in the field of NMR. In both cases the loss of polarization (magnetization) due to dephasing of the spins in time is restored by an effective time reversal operation, that leads to a restituation of polarization (rephasing). In NMR the dephasing happens due to variation in the local fields at positions of the nuclei, in NSE the dephasing is due to different neutron velocities in the incoming neutron beam. The Larmor precession of the neutron spin in a preparation zone with a magnetic field before the sample encodes the individual velocities of neutrons in the beam into precession angles. Close to the sample the time reversal is effected by a so-called flipper. A symmetric decoding zone follows such that at its end the precession angle accumulated in the preparation zone is exactly compensated (provided the sample did not change the neutron velocity, i.e. elastic scattering), all spins rephase to form the "spin-echo". Ideally the full polarization is restored. This effect does not depend on the velocity/energy/wavelength of the incoming neutron. If the scattering at the sample is not elastic but changes the neutron velocity, the rephasing will become incomplete and a loss of final polarization results, which depends on the distribution of differences in the time, which the neutrons need to fly through the symmetric first (coding) and second (decoding)precession zones. The time differences occur due to a velocity change acquired by non-elastic scattering at the sample. The distribution of these time differences is proportional (in the linearization approximation which is appropriate for quasi-elastic high resolution spectroscopy) to the spectral part of the scattering function S(Q,ω). The effect on the measured beam polarization is proportional to the cos-Fourier transform of the spectral function, the intermediate scattering function F(Q,t). The time parameter depends on the neutron wavelength and the factor connecting precession angle with (reciprocal) velocity, which can e.g. be controlled by setting a certain magnetic field in the preparation and decoding zones. Scans of t may then be performed by varying the magnetic field.

It is important to note: that all the spin manipulations are just a means to detect velocity changes of the neutron, which influence—for technical reasons—in terms of a Fourier transform of the spectral function in the measured intensity. The velocity changes of the neutrons convey the physical information which is available by using NSE, i.e.

$$I(Q,t) \propto S(Q) + \int \cos(\omega t) S(Q,\omega) dt \text{ where } \omega \propto \Delta v \text{ and } t \propto B \times \lambda^3.$$

B denotes the precession field strength, λ the (average) neutron wavelength and Δv the neutron velocity change upon scattering at the sample.

The main reason for using NSE is that by the above means it can reach Fourier times of up to many 100ns, which corresponds to energy resolutions in the neV range. The closest approach to this resolution by a spectroscopic neutron instrument type, namely the backscattering spectrometer (BSS), is in the range of 0.5 to 1 μeV. The spin-echo trick allows to use an intense beam of neutrons with a wavelength distribution of 10% or more and at the same time to be sensitive to velocity changes in the range of less than 10^{-4}.

Note: the above explanations assumes the generic NSE configuration—as first utilized by the IN11 instrument at the Institut Laue–Langevin (ILL)--. Other approaches are possible like the resonance spin-echo, NRSE with concentrated a DC field and a RF field in the flippers at the end of preparation and decoding zones which then are without magnetic field (zero field). In principle these approaches are equivalent concerning the connection of the final intensity signal with the intermediate scattering function. Due to technical difficulties until now they have not reached the same level of performance than the generic (IN11) NSE types.

What it Can Measure

In soft matter research the structure of macromolecular objects is often investigated by small angle neutron scattering, SANS. The exchange of hydrogen with deuterium in some of the molecules creates scattering contrast between even equal chemical species. The SANS diffraction pattern—if interpreted in real space—corresponds to a snapshot picture of the molecular arrangement. Neutron spin echo instruments can analyze the inelastic broadening of the SANS intensity and thereby analyze the motion of the macromolecular objects. A coarse analogy would be a photo with a certain opening time instead of the SANS like snapshot(So we can analyze the vibration frequency of the molecules as well as arrangement). The opening time corresponds to the Fourier time which depends on the setting of the NSE spectrometer, it is proportional to the magnetic field (integral) and to the third power of the neutron wavelength. Values up to several hundreds of nanoseconds are available. Note that the spatial resolution of the scattering experiment is in the nanometer range, which means that a time range of e.g. 100 ns corresponds to effective molecular motion velocities of 1 nm/100 ns = 1 cm/s. This may be compared to the typical neutron velocity of 200..1000 m/s used in these type of experiments.

NSE and Spin-incoherent Scattering (from Protons)

Many inelastic studies that use normal time-of-flight (TOF) or backscattering spectrometers rely on the huge incoherent neutron scattering cross section of protons. The scattering signal is dominated by the corresponding contribution, which represents the (average) self-correlation function (in time) of the protons.

For NSE spin incoherent scattering has the disadvantage that it flips the neutron spins during scattering with a probability of 2/3. Thus converting 2/3 of the scattering intensity into "non-polarized" background and putting a factor of -1/3 in front of the cos-Fourier integral contribution pertaining the incoherent intensity. This signal subtracts from the coherent echo signal. The result may be a complicated combination which cannot be decomposed if only NSE is employed. However, in pure cases, i.e. when there is an overwhelming intensity contribution due to protons, NSE can be used to measure their incoherent spectrum.

The intensity situation of NSE—for e.g. soft-matter samples—is the same as in small angle scattering (SANS). Which means that molecular objects with coherent scattering contrast at low Q (momentum transfer) show a much larger intensity as the incoherent contribution (which is the background level). But at larger Q usually somewhere around Q=0.3 A^{-1} the incoherent scattering becomes stronger than the coherent part. At least for hydrogen containing systems contrast requires the presence of some protons and even pure deuterated samples show spin-incoherent scattering from deuterons, however, 40 times weaker than the proton scattering.

Fully protonated samples allow successful measurements but at intensities of the order of the SANS background level. This requires correspondingly long counting times.

Note: This interference with the spin manipulation of the NSE technique occurs only with spin-incoherent scattering. Isotopic incoherent scattering yields a "normal" NSE signal.

Existing Spectrometers

IN11 (ILL, Grenoble, France)

IN15 (ILL, Grenoble, France)

NL2a J-NSE (JCNS, Juelich, Germany, hosted by FRMII, Munich (Garching), Germany)

NL5-S RESEDA (FRM II Munich, Munich, Germany)

V5/SPAN (Hahn-Meitner Institut, Berlin, Germany)

C2-3-1 NSE ISSP, (JRR-3, Tokai, Japan).

BL-15 NSE (SNS, ORNL, Oak Ridge, USA)

NG5-NSE (CHRNS, NIST, Gaithersburg, USA),

Neuroimaging

Neuroimaging or brain imaging is the use of various techniques to either directly or indirectly image the structure, function/pharmacology of the nervous system. It is a relatively new discipline within medicine, neuroscience, and psychology. Physicians who specialize in the performance and interpretation of neuroimaging in the clinical setting are neuroradiologists.

Neuroimaging falls into two broad categories:

- Structural imaging, which deals with the structure of the nervous system and the diagnosis of gross (large scale) intracranial disease (such as tumor), and injury, and

- Functional imaging, which is used to diagnose metabolic diseases and lesions on a finer scale (such as Alzheimer's disease) and also for neurological and cognitive psychology research and building brain-computer interfaces.

Functional imaging enables, for example, the processing of information by centers in the brain to be visualized directly. Such processing causes the involved area of the brain to increase metabolism and "light up" on the scan. One of the more controversial uses of neuroimaging has been research into "thought identification" or mind-reading.

History

The first chapter of the history of neuroimaging traces back to the Italian neuroscientist Angelo Mosso who invented the 'human circulation balance', which could non-invasively measure the redistribution of blood during emotional and intellectual activity. However, even if only briefly mentioned by William James in 1890, the details and precise workings of this balance and the experiments Mosso performed with it have remained largely unknown until the recent discovery of the original instrument as well as Mosso's reports by Stefano Sandrone and colleagues.

In 1918 the American neurosurgeon Walter Dandy introduced the technique of ventriculography. X-ray images of the ventricular system within the brain were obtained by injection of filtered air directly into one or both lateral ventricles of the brain. Dandy also observed that air introduced into the subarachnoid space via lumbar spinal puncture could enter the cerebral ventricles and also demonstrate the cerebrospinal fluid compartments around the base of the brain and over its surface. This technique was called pneumoencephalography.

In 1927 Egas Moniz introduced cerebral angiography, whereby both normal and abnormal blood vessels in and around the brain could be visualized with great precision.

In the early 1970s, Allan McLeod Cormack and Godfrey Newbold Hounsfield introduced computerized axial tomography (CAT or CT scanning), and ever more detailed anatomic images of the brain became available for diagnostic and research purposes. Cormack and Hounsfield won the 1979 Nobel Prize for Physiology or Medicine for their work. Soon after the introduction of CAT in the early 1980s, the development of radioligands allowed single photon emission computed tomography (SPECT) and positron emission tomography (PET) of the brain.

More or less concurrently, magnetic resonance imaging (MRI or MR scanning) was developed by researchers including Peter Mansfield and Paul Lauterbur, who were awarded the Nobel Prize for Physiology or Medicine in 2003. In the early 1980s MRI was introduced clinically, and during the 1980s a veritable explosion of technical refinements and diagnostic MR applications took place. Scientists soon learned that the large blood flow changes measured by PET could also be imaged by the correct type of MRI. Functional magnetic resonance imaging (fMRI) was born, and since the 1990s, fMRI has come to dominate the brain mapping field due to its low invasiveness, lack of radiation exposure, and relatively wide availability.

In the early 2000s the field of neuroimaging reached the stage where limited practical applications of functional brain imaging have become feasible. The main application area is crude forms of brain-computer interface.

Indications

Neuroimaging follows a neurological examination in which a physician has found cause to more deeply investigate a patient who has or may have a neurological disorder.

One of the more common neurological problems which a person may experience is simple syncope. In cases of simple syncope in which the patient's history does not suggest other neurological symptoms, the diagnosis includes a neurological examination but routine neurological imaging is not indicated because the likelihood of finding a cause in the central nervous system is extremely low and the patient is unlikely to benefit from the procedure.

Neuroimaging is not indicated for patients with stable headaches which are diagnosed as migraine. Studies indicate that presence of migraine does not increase a patient's risk for intracranial disease. A diagnosis of migraine which notes the absence of other problems, such as papilledema, would not indicate a need for neuroimaging. In the course of conducting a careful diagnosis, the physician should consider whether the headache has a cause other than the migraine and might require neuroimaging.

Another indication for neuroimaging is CT-, MRI- and PET-guidedstereotactic surgery or radiosurgery for treatment of intracranial tumors, arteriovenous malformations and other surgically treatable conditions.

Brain Imaging Techniques
Computed Axial Tomography

Computed tomography (CT) or *Computed Axial Tomography* (CAT) scanning uses a series of x-rays of the head taken from many different directions. Typically used for quickly viewing brain injuries, CT scanning uses a computer program that performs a numerical integral calculation (the inverse Radon transform) on the measured x-ray series to estimate how much of an x-ray beam is absorbed in a small volume of the brain. Typically the information is presented as cross sections of the brain.

Diffuse Optical Imaging

Diffuse optical imaging (DOI) or diffuse optical tomography (DOT) is a medical imaging modality which uses near infrared light to generate images of the body. The technique measures the optical absorption of haemoglobin, and relies on the absorption spectrum of haemoglobin varying with its oxygenation status. High-density diffuse optical tomography (HD-DOT) has seen setbacks due to limited resolution. Early results have been promising, a comparison and validation of diffuse optical imaging against the standard of functional magnetic resonance imaging (fMRI) has been lacking. HD-DOT has adequate image quality to be useful as a surrogate for fMRI.

Event-related Optical Signal

Event-related optical signal (EROS) is a brain-scanning technique which uses infrared light through optical fibers to measure changes in optical properties of active areas of the cerebral cortex. Whereas techniques such as diffuse optical imaging (DOT) and near infrared spectroscopy (NIRS) measure optical absorption of haemoglobin, and thus are based on blood flow, EROS takes advantage of the scattering properties of the neurons themselves, and thus provides a much more direct measure of cellular activity. EROS can pinpoint activity in the brain within millimeters (spatially) and within milliseconds (temporally). Its biggest downside is the inability to detect activity more than a few centimeters deep. EROS is a new, relatively inexpensive technique that is non-invasive to the test subject. It was developed at the University of Illinois at Urbana-Champaign where it is now used in the Cognitive Neuroimaging Laboratory of Dr. Gabriele Gratton and Dr. Monica Fabiani.

Magnetic Resonance Imaging

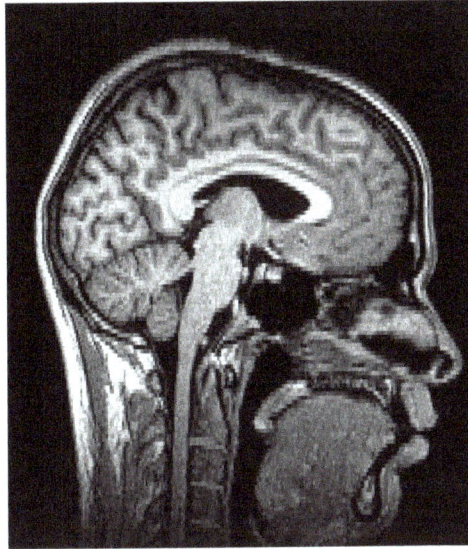

Sagittal MRI slice at the midline.

Magnetic resonance imaging (MRI) uses magnetic fields and radio waves to produce high quality two- or three-dimensional images of brain structures without use of ionizing radiation (X-rays) or radioactive tracers.

Functional Magnetic Resonance Imaging

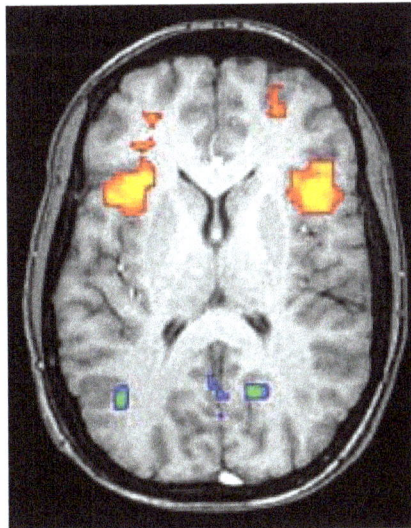

Axial MRI slice at the level of the basal ganglia, showing fMRI BOLD
signal changes overlaid in red (increase) and blue (decrease) tones.

Functional magnetic resonance imaging (fMRI) and arterial spin labeling (ASL) relies on the paramagnetic properties of oxygenated and deoxygenated hemoglobin to see images of changing blood flow in the brain associated with neural activity. This allows images to be generated that reflect which brain structures are activated (and how) during performance of different tasks or at resting state. According to the oxygenation hypothesis, changes in oxygen usage in regional cerebral blood flow during cognitive or behavioral activity can be associated with the regional neurons as being directly related to the cognitive or behavioral tasks being attended.

Most fMRI scanners allow subjects to be presented with different visual images, sounds and touch stimuli, and to make different actions such as pressing a button or moving a joystick. Consequently, fMRI can be used to reveal brain structures and processes associated with perception, thought and action. The resolution of fMRI is about 2-3 millimeters at present, limited by the spatial spread of the hemodynamic response to neural activity. It has largely superseded PET for the study of brain activation patterns. PET, however, retains the significant advantage of being able to identify specific brain receptors (or transporters) associated with particular neurotransmitters through its ability to image radiolabelled receptor "ligands" (receptor ligands are any chemicals that stick to receptors).

As well as research on healthy subjects, fMRI is increasingly used for the medical diagnosis of disease. Because fMRI is exquisitely sensitive to oxygen usage in blood flow, it is extremely sensitive to early changes in the brain resulting from ischemia (abnormally low blood flow), such as the changes which follow stroke. Early diagnosis of certain types of stroke is increasingly important in neurology, since substances which dissolve blood clots may be used in the first few hours after certain types of stroke occur, but are dangerous to use afterwards. Brain changes seen on fMRI may help to make the decision to treat with these agents. With between 72% and 90% accuracy where chance would achieve 0.8%, fMRI techniques can decide which of a set of known images the subject is viewing.

Magnetoencephalography

Magnetoencephalography (MEG) is an imaging technique used to measure the magnetic fields produced by electrical activity in the brain via extremely sensitive devices such as superconducting quantum interference devices (SQUIDs). MEG offers a very direct measurement of neural electrical activity (compared to fMRI for example) with very high temporal resolution but relatively low spatial resolution. The advantage of measuring the magnetic fields produced by neural activity is that they are likely to be less distorted by surrounding tissue (particularly the skull and scalp) compared to the electric fields measured by electroencephalography (EEG). Specifically, it can be shown that magnetic fields produced by electrical activity are not affected by the surrounding head tissue, when the head is modeled as a set of concentric spherical shells, each being an isotropic homogeneous conductor. Real heads are non-spherical and have largely anisotropic conductivities (particularly white matter and skull). While skull anisotropy has negligible effect on MEG (unlike EEG), white matter anisotropy strongly affects MEG measurements for radial and deep sources. Note, however, that the skull was assumed to be uniformly anisotropic in this study, which is not true for a real head: the absolute and relative thicknesses of diploë and tables layers vary among and within the skull bones. This makes it likely that MEG is also affected by the skull anisotropy, although probably not to the same degree as EEG.

There are many uses for MEG, including assisting surgeons in localizing a pathology, assisting researchers in determining the function of various parts of the brain, neuro-feedback, and others.

Positron Emission Tomography

Positron emission tomography (PET) measures emissions from radioactively labeled metabolically active chemicals that have been injected into the bloodstream. The emission data are computer-processed to produce 2- or 3-dimensional images of the distribution of the chemicals throughout the brain. The positron emitting radioisotopes used are produced by a cyclotron, and chemicals are labeled with these radioactive atoms. The labeled compound, called a *radiotracer*, is injected into the bloodstream and eventually makes its way to the brain. Sensors in the PET scanner detect the radioactivity as the compound accumulates in various regions of the brain. A computer uses the data gathered by the sensors to create multicolored 2- or 3-dimensional images that show where the compound acts in the brain. Especially useful are a wide array of ligands used to map different aspects of neurotransmitter activity, with by far the most commonly used PET tracer being a labeled form of glucose.

The greatest benefit of PET scanning is that different compounds can show blood flow and oxygen and glucosemetabolism in the tissues of the working brain. These measurements reflect the amount of brain activity in the various regions of the brain and allow to learn more about how the brain works. PET scans were superior to all other metabolic imaging methods in terms of resolution and speed of completion (as little as 30 seconds), when they first became available. The improved resolution permitted better study to be made as to the area of the brain activated by a particular task. The biggest drawback of PET scanning is that because the radioactivity decays rapidly, it is limited to monitoring short tasks. Before fMRI technology came online, PET scanning was the preferred method of functional (as opposed to structural) brain imaging, and it continues to make large contributions to neuroscience.

PET scanning is also used for diagnosis of brain disease, most notably because brain tumors, strokes, and neuron-damaging diseases which cause dementia (such as Alzheimer's disease) all cause great changes in brain metabolism, which in turn causes easily detectable changes in PET scans. PET is probably most useful in early cases of certain dementias (with classic examples being Alzheimer's disease and Pick's disease) where the early damage is too diffuse and makes too little difference in brain volume and gross structure to change CT and standard MRI images enough to be able to reliably differentiate it from the "normal" range of cortical atrophy which occurs with aging (in many but not all) persons, and which does *not* cause clinical dementia.

Single-photon Emission Computed Tomography

Single-photon emission computed tomography (SPECT) is similar to PET and uses gamma ray-emitting radioisotopes and a gamma camera to record data that a computer uses to construct two- or three-dimensional images of active brain regions. SPECT relies on an injection of radioactive tracer, or "SPECT agent," which is rapidly taken up by the brain but does not redistribute. Uptake of SPECT agent is nearly 100% complete within 30 to 60 seconds, reflecting cerebral blood flow (CBF) at the time of injection. These properties of SPECT make it particularly well-suited for epilepsy imaging, which is usually made difficult by problems with patient movement and variable seizure types. SPECT provides a "snapshot" of cerebral blood flow since scans can be acquired after seizure termination (so long as the radioactive tracer was injected at the time of the seizure). A significant limitation of SPECT is its poor resolution (about 1 cm) compared to that of MRI. Today, SPECT machines with Dual Detector Heads are commonly used, although Triple Detector Head machines are available in the marketplace. Tomographic reconstruction, (mainly used for functional "snapshots" of the brain) requires multiple projections from Detector Heads which rotate around the human skull, so some researchers have developed 6 and 11 Detector Head SPECT machines to cut imaging time and give higher resolution.

Like PET, SPECT also can be used to differentiate different kinds of disease processes which produce dementia, and it is increasingly used for this purpose. Neuro-PET has a disadvantage of requiring use of tracers with half-lives of at most 110 minutes, such as FDG. These must be made in a cyclotron, and are expensive or even unavailable if necessary transport times are prolonged more than a few half-lives. SPECT, however, is able to make use of tracers with much longer half-lives, such as technetium-99m, and as a result, is far more widely available.

Cranial Ultrasound

Cranial ultrasound is usually only used in babies, whose open fontanelles provide acoustic windows allowing ultrasound imaging of the brain. Advantages include absence of ionising radiation and the possibility of bedside scanning, but the lack of soft-tissue detail means MRI may be preferred for some conditions.

Comparison of Imaging Types

Magnetic Resonance Imaging (MRI) depends on magnetic activity in the brain and does not use X-rays, so it is considered more safe than imaging techniques that do use X-rays. SPECT uses gamma rays, which are characteristically more safe than other imaging systems using alpha or beta rays. Both PET and SPECT scans require the injection of radioactive materials, but the half-lives of isotopes used in SPECT can be more easily managed.

Criticism and Cautions

Some scientists have criticized the brain image-based claims made in scientific journals and the popular press, like the discovery of "the part of the brain responsible" things like love or musical ability or a specific memory. Many mapping techniques have a relatively low resolution, including hundreds of thousands of neurons in a single voxel. Many functions also involve multiple parts of the brain, meaning that this type of claim is probably both unverifiable with the equipment used, and generally based on an incorrect assumption about how brain functions are divided. It may be that most brain functions will only be described correctly after being measured with much more fine-grained measurements that look not at large regions but instead at a very large number of tiny individual brain circuits. Many of these studies also have technical problems like small sample size or poor equipment calibration which means they cannot be reproduced - considerations which are sometimes ignored to produce a sensational journal article or news headline. In some cases the brain mapping techniques are used for commercial purposes, lie detection, or medical diagnosis in ways which have not been scientifically validated.

Calorimetry

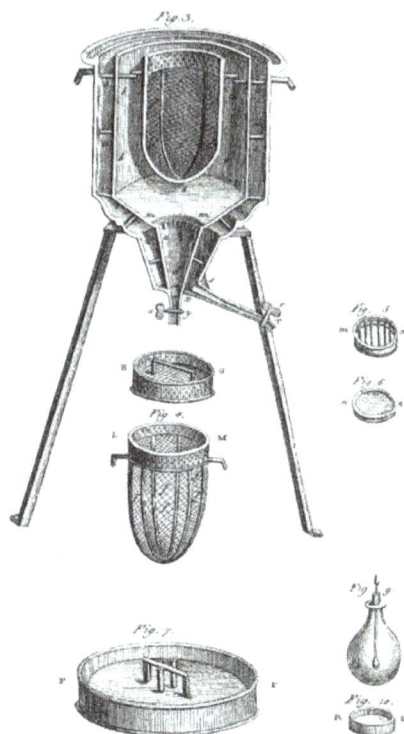

The world's first ice-calorimeter, used in the winter of 1782–83, by Antoine Lavoisier and Pierre-Simon Laplace, to determine the heat involved in various chemical changes; calculations which were based on Joseph Black's prior discovery of latent heat. These experiments mark the foundation of thermochemistry.

Snellen direct calorimetry chamber, University of Ottawa.

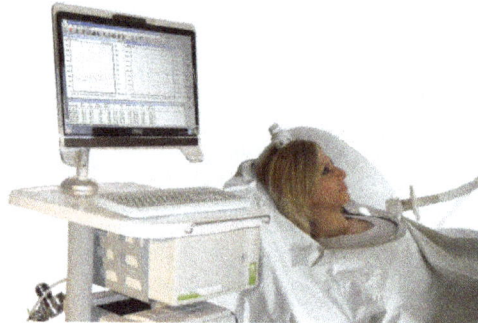

Indirect calorimetry metabolic cart measuring oxygen uptake and CO2 production of a spontaneously breathing subject (dilution method with canopy hood).

Calorimetry is the science or act of measuring changes in state variables of a body for the purpose of deriving the heat transfer associated with changes of its state due for example to chemical reactions, physical changes, or phase transitions under specified constraints. Calorimetry is performed with a calorimeter. The word *calorimetry* is derived from the Latin word *calor*, meaning heat and the Greek word (metron), meaning measure. Scottish physician and scientist Joseph Black, who was the first to recognize the distinction between heat and temperature, is said to be the founder of the science of calorimetry.

Indirect Calorimetry calculates heat that living organisms produce by measuring either their production of carbon dioxide and nitrogen waste (frequently ammonia in aquatic organisms, or urea in terrestrial ones), or from their consumption of oxygen. Lavoisier noted in 1780 that heat production can be predicted from oxygen consumption this way, using multiple regression. The Dynamic Energy Budget theory explains why this procedure is correct. Heat generated by living organisms may also be measured by direct calorimetry, in which the entire organism is placed inside the calorimeter for the measurement.

A widely used modern instrument is the differential scanning calorimeter, a device which allows thermal data to be obtained on small amounts of material. It involves heating the sample at a controlled rate and recording the heat flow either into or from the specimen.

Classical Calorimetric Calculation of Heat

Basic Classical Calculation with Respect to Volume

Calorimetry requires that a reference material that changes temperature have known definite thermal constitutive properties. The classical rule, recognized by Clausius and by Kelvin, is that the pressure exerted by the calorimetric material is fully and rapidly determined solely by its temperature and volume; this rule is for changes that do not involve phase change, such as melting of ice. There are many materials that do not comply with this rule, and for them, the present formula of classical calorimetry does not provide an adequate account. Here the classical rule is assumed to hold for the calorimetric material being used, and the propositions are mathematically written:

The thermal response of the calorimetric material is fully described by its pressure p as the value of its constitutive function $p(V,T)$ of just the volume V and the temperature T. All increments are here required to be very small. This calculation refers to a domain of volume and temperature of the body in which no phase change occurs, and there is only one phase present. An important assumption here is continuity of property relations. A different analysis is needed for phase change

When a small increment of heat is gained by a calorimetric body, with small increments, δV of its volume, and δT of its temperature, the increment of heat, δQ, gained by the body of calorimetric material, is given by

$$\delta Q = C_T^{(V)}(V,T)\delta V + C_V^{(T)}(V,T)\delta T$$

where

$C_T^{(V)}(V,T)$ denotes the latent heat with respect to volume, of the calorimetric material at constant controlled temperature T. The surroundings' pressure on the material is instrumentally adjusted to impose a chosen volume change, with initial volume V. To determine this latent heat, the volume change is effectively the independently instrumentally varied quantity. This latent heat is not one of the widely used ones, but is of theoretical or conceptual interest.

$C_V^{(T)}(V,T)$ denotes the heat capacity, of the calorimetric material at fixed constant volume V , while the pressure of the material is allowed to vary freely, with initial temperature T . The temperature is forced to change by exposure to a suitable heat bath. It is customary to write $C_V^{(T)}(V,T)$ simply as $C_V(V,T)$, or even more briefly as C_V . This latent heat is one of the two widely used ones.

The latent heat with respect to volume is the heat required for unit increment in volume at constant temperature. It can be said to be 'measured along an isotherm', and the pressure the material exerts is allowed to vary freely, according to its constitutive law $p = p(V,T)$.. For a given material, it can have a positive or negative sign or exceptionally it can be zero, and this can depend on the temperature, as it does for water about 4 C. The concept of latent heat with respect to volume was perhaps first recognized by Joseph Black in 1762. The term 'latent heat of expansion' is also used. The latent heat with respect to volume can also be called the 'latent energy with respect to volume'. For all of these usages of 'latent heat', a more systematic terminology uses 'latent heat capacity'.

The heat capacity at constant volume is the heat required for unit increment in temperature at constant volume. It can be said to be 'measured along an isochor', and again, the pressure the material exerts is allowed to vary freely. It always has a positive sign. This means that for an increase in the temperature of a body without change of its volume, heat must be supplied to it. This is consistent with common experience.

Quantities like δQ are sometimes called 'curve differentials', because they are measured along curves in the (V,T) surface.

Classical Theory for Constant-volume Calorimetry

Constant-volume calorimetry is calorimetry performed at a constant volume. This involves the use of a constant-volume calorimeter. Heat is still measured by the above-stated principle of calorimetry.

This means that in a suitably constructed calorimeter, called a bomb calorimeter, the increment of volume δV can be made to vanish, $\delta V = 0$. For constant-volume calorimetry:

$$\delta Q = C_V \delta T$$

where

δT denotes the increment in temperature and

C_V denotes the heat capacity at constant volume.

Classical Heat Calculation with Respect to Pressure

From the above rule of calculation of heat with respect to volume, there follows one with respect to pressure.

In a process of small increments, δp of its pressure, and δT of its temperature, the increment of heat, δQ, gained by the body of calorimetric material, is given by

$$\delta Q = C_T^{(p)}(p,T)\delta p + C_p^{(T)}(p,T)\delta T$$

where

$C_T^{(p)}(p,T)$ denotes the latent heat with respect to pressure, of the calorimetric material at constant temperature, while the volume and pressure of the body are allowed to vary freely, at pressure p and temperature T ;

$C_p^{(T)}(p,T)$ denotes the heat capacity, of the calorimetric material at constant pressure, while the temperature and volume of the body are allowed to vary freely, at pressure p and temperature T. It is customary to write $C_p^{(T)}(p,T)$ simply as $C_p(p,T)$, or even more briefly as C_p .

The new quantities here are related to the previous ones:

$$C_T^{(p)}(p,T) = \frac{C_T^{(V)}(V,T)}{\left.\dfrac{\partial p}{\partial V}\right|_{(V,T)}}$$

$$C_p^{(T)}(p,T) = C_V^{(T)}(V,T) - C_T^{(V)}(V,T)\frac{\left.\dfrac{\partial p}{\partial T}\right|_{(V,T)}}{\left.\dfrac{\partial p}{\partial V}\right|_{(V,T)}}$$

where

$\left.\dfrac{\partial p}{\partial V}\right|_{(V,T)}$ denotes the partial derivative of $p(V,T)$ with respect to V evaluated

for (V,T)

and

$\left.\dfrac{\partial p}{\partial T}\right|_{(V,T)}$ denotes the partial derivative of $p(V,T)$ with respect to T evaluated for

(V,T).

The latent heats $C_T^{(V)}(V,T)$ and $C_T^{(p)}(p,T)$ are always of opposite sign.

It is common to refer to the ratio of specific heats as

$$\gamma(V,T) = \frac{C_p^{(T)}(p,T)}{C_V^{(T)}(V,T)} \text{ often just written as } \gamma = \frac{C_p}{C_V}.$$

Calorimetry Through Phase Change, Equation of State Shows One Jump Discontinuity

An early calorimeter was that used by Laplace and Lavoisier, as shown in the figure above. It worked at constant temperature, and at atmospheric pressure. The latent heat involved was then not a latent heat with respect to volume or with respect to pressure, as in the above account for calorimetry without phase change. The latent heat involved in this calorimeter was with respect to phase change, naturally occurring at constant temperature. This kind of calorimeter worked by measurement of mass of water produced by the melting of ice, which is a phase change.

Cumulation of Heating

For a time-dependent process of heating of the calorimetric material, defined by a continuous joint progression $P(t_1, t_2)$ of $V(t)$ and $T(t)$, starting at time t_1 and ending at time t_2, there can be calculated an accumulated quantity of heat delivered, $\Delta Q(P(t_1, t_2))$. This calculation is done by mathematical integration along the progression with respect to time. This is because increments of heat are 'additive'; but this does not mean that heat is a conservative quantity. The idea that heat was a conservative quantity was invented by Lavoisier, and is called the 'caloric theory'; by the middle of the nineteenth century it was recognized as mistaken. Written with the symbol Ä , the quantity $\Delta Q(P(t_1, t_2))$ is not at all restricted to be an increment with very small values; this is in contrast with δQ .

One can write

$$\Delta Q(P(t_1, t_2))$$

$$= \int_{P(t_1, t_2)} \dot{Q}(t)dt$$

$$= \int_{P(t_1, t_2)} C_T^{(V)}(V,T)\dot{V}(t)dt + \int_{P(t_1, t_2)} C_V^{(T)}(V,T)\dot{T}(t)dt .$$

This expression uses quantities such as $\dot{Q}(t)$ which are defined in the section below headed 'Mathematical aspects of the above rules'.

Mathematical Aspects of the Above Rules

The use of 'very small' quantities such as δQ is related to the physical requirement for the quantity $p(V,T)$ to be 'rapidly determined' by V and T ; such 'rapid determination' refers to a physical process. These 'very small' quantities are used in the Leibniz approach to the infinitesimal calculus. The Newton approach uses instead 'fluxions' such as $\dot{V}(t) = \dfrac{dV}{dt}\bigg|_{t}$, which makes it more obvious that $p(V,T)$ must be 'rapidly determined'.

In terms of fluxions, the above first rule of calculation can be written

$$\dot{Q}(t) = C_T^{(V)}(V,T)\dot{V}(t) + C_V^{(T)}(V,T)\dot{T}(t)$$

where

  t denotes the time

  $\dot{Q}(t)$ denotes the time rate of heating of the calorimetric material at time t

  $\dot{V}(t)$ denotes the time rate of change of volume of the calorimetric material at time t

  $\dot{T}(t)$ denotes the time rate of change of temperature of the calorimetric material.

The increment δQ and the fluxion $\dot{Q}(t)$ are obtained for a particular time t that determines the values of the quantities on the righthand sides of the above rules. But this is not a reason to expect that there should exist a mathematical function $Q(V,T)$. For this reason, the increment δQ is said to be an 'imperfect differential' or an 'inexact differential'. Some books indicate this by writing q instead of δQ. Also, the notation $đQ$ is used in some books. Carelessness about this can lead to error.

The quantity $\ddot{A}Q(P(t_1,t_2))$ is properly said to be a functional of the continuous joint progression $P(t_1,t_2)$ of $V(t)$ and $T(t)$, but, in the mathematical definition of a function, $\Delta Q(P(t_1,t_2))$ is not a function of (V,T). Although the fluxion $\dot{Q}(t)$ is defined here as a function of time t, the symbols Q and $Q(V,T)$ respectively standing alone are not defined here.

Physical Scope of the Above Rules of Calorimetry

The above rules refer only to suitable calorimetric materials. The terms 'rapidly' and 'very small' call for empirical physical checking of the domain of validity of the above rules.

The above rules for the calculation of heat belong to pure calorimetry. They make no reference to thermodynamics, and were mostly understood before the advent of thermodynamics. They are the basis of the 'thermo' contribution to thermodynamics. The 'dynamics' contribution is based on the idea of work, which is not used in the above rules of calculation.

Experimentally Conveniently Measured Coefficients

Empirically, it is convenient to measure properties of calorimetric materials under experimentally controlled conditions.

Pressure Increase at Constant Volume

For measurements at experimentally controlled volume, one can use the assumption, stated above, that the pressure of the body of calorimetric material is can be expressed as a function of its volume and temperature.

For measurement at constant experimentally controlled volume, the isochoric coefficient of pressure rise with temperature, is defined by

$$\alpha_V(V,T) = \frac{\left.\dfrac{\partial p}{\partial V}\right|_{(V,T)}}{p(V,T)}.$$

Expansion at Constant Pressure

For measurements at experimentally controlled pressure, it is assumed that the volume V of the body of calorimetric material can be expressed as a function $V(T,p)$ of its temperature T and pressure p. This assumption is related to, but is not the same as, the above used assumption that the pressure of the body of calorimetric material is known as a function of its volume and temperature; anomalous behaviour of materials can affect this relation.

The quantity that is conveniently measured at constant experimentally controlled pressure, the isobaric volume expansion coefficient, is defined by

$$\beta_p(T,p) = \frac{\left.\dfrac{\partial V}{\partial T}\right|_{(T,p)}}{V(T,p)}.$$

Compressibility at Constant Temperature

For measurements at experimentally controlled temperature, it is again assumed that the volume V of the body of calorimetric material can be expressed as a function $V(T,p)$ of its temperature T and pressure p, with the same provisos as mentioned just above.

The quantity that is conveniently measured at constant experimentally controlled temperature, the isothermal compressibility, is defined by

$$\kappa_T(T,p) = -\frac{\left.\frac{\partial V}{\partial p}\right|_{(T,p)}}{V(T,p)}.$$

Relation between Classical Calorimetric Quantities

Assuming that the rule $p = p(V,T)$ is known, one can derive the function $\frac{\partial p}{\partial T}$ that is used above in the classical heat calculation with respect to pressure. This function can be found experimentally from the coefficients $\beta_p(T,p)$ and $\kappa_T(T,p)$ through the mathematically deducible relation

$$\frac{\partial p}{\partial T} = \frac{\beta_p(T,p)}{\kappa_T(T,p)}.$$

Connection between Calorimetry and Thermodynamics

Thermodynamics developed gradually over the first half of the nineteenth century, building on the above theory of calorimetry which had been worked out before it, and on other discoveries. According to Gislason and Craig (2005): "Most thermodynamic data come from calorimetry..." According to Kondepudi (2008): "Calorimetry is widely used in present day laboratories."

In terms of thermodynamics, the internal energy U of the calorimetric material can be considered as the value of a function $U(V,T)$ of (V,T), with partial derivatives $\frac{\partial U}{\partial V}$ and $\frac{\partial U}{\partial T}$.

Then it can be shown that one can write a thermodynamic version of the above calorimetric rules:

$$\delta Q = \left[p(V,T) + \left.\frac{\partial U}{\partial V}\right|_{(V,T)} \right]\delta V + \left.\frac{\partial U}{\partial T}\right|_{(V,T)} \delta T$$

with

$$C_T^{(V)}(V,T) = p(V,T) + \left.\frac{\partial U}{\partial V}\right|_{(V,T)}$$

and

$$C_V^{(T)}(V,T) = \left.\frac{\partial U}{\partial T}\right|_{(V,T)}.$$

Again, further in terms of thermodynamics, the internal energy U of the calorimetric material can sometimes, depending on the calorimetric material, be considered as the value of a function $U(p,T)$ of (p,T), with partial derivatives $\dfrac{\partial U}{\partial p}$ and $\dfrac{\partial U}{\partial T}$, and with V being expressible as the value of a function $V(p,T)$ of (p,T), with partial derivatives $\dfrac{\partial V}{\partial p}$ and $\dfrac{\partial V}{\partial T}$.

Then, according to Adkins (1975), it can be shown that one can write a further thermodynamic version of the above calorimetric rules:

$$\delta Q = \left[\left. \frac{\partial U}{\partial p} \right|_{(p,T)} + p \left. \frac{\partial V}{\partial p} \right|_{(p,T)} \right] \delta p + \left[\left. \frac{\partial U}{\partial T} \right|_{(p,T)} + p \left. \frac{\partial V}{\partial T} \right|_{(p,T)} \right] \delta T$$

with

$$C_T^{(p)}(p,T) = \left. \frac{\partial U}{\partial p} \right|_{(p,T)} + p \left. \frac{\partial V}{\partial p} \right|_{(p,T)}$$

and

$$C_p^{(T)}(p,T) = \left. \frac{\partial U}{\partial T} \right|_{(p,T)} + p \left. \frac{\partial V}{\partial T} \right|_{(p,T)} .$$

Beyond the calorimetric fact noted above that the latent heats $C_T^{(V)}(V,T)$ and $C_T^{(p)}(p,T)$ are always of opposite sign, it may be shown, using the thermodynamic concept of work, that also

$$C_T^{(V)}(V,T) \left. \frac{\partial p}{\partial T} \right|_{(V,T)} \geq 0.$$

Special Interest of Thermodynamics in Calorimetry: the Isothermal Segments of a Carnot Cycle

Calorimetry has a special benefit for thermodynamics. It tells about the heat absorbed or emitted in the isothermal segment of a Carnot cycle.

A Carnot cycle is a special kind of cyclic process affecting a body composed of material suitable for use in a heat engine. Such a material is of the kind considered in calorimetry, as noted above, that exerts a pressure that is very rapidly determined just by temperature and volume. Such a body is said to change reversibly. A Carnot cycle consists of four successive stages or segments:

(1) a change in volume from a volume V_a to a volume V_b at constant temperature T^+ so as to incur a flow of heat into the body (known as an isothermal change)

(2) a change in volume from V_b to a volume V_c at a variable temperature just such as to incur no flow of heat (known as an adiabatic change)

(3) another isothermal change in volume from V_c to a volume V_d at constant temperature T^- such as to incur a flow or heat out of the body and just such as to precisely prepare for the following change

(4) another adiabatic change of volume from V_d back to V_a just such as to return the body to its starting temperature T^+ .

In isothermal segment (1), the heat that flows into the body is given by

$$\Delta Q(V_a,V_b;T^+)= \int_{V_a}^{V_b} C_T^{(V)}(V,T^+)dV$$

and in isothermal segment (3) the heat that flows out of the body is given by

$$-\Delta Q(V_c,V_d;T^-)=-\int_{V_c}^{V_d} C_T^{(V)}(V,T^-)dV .$$

Because the segments (2) and (4) are adiabats, no heat flows into or out of the body during them, and consequently the net heat supplied to the body during the cycle is given by

$$\Delta Q(V_a,V_b;T^+;V_c,V_d;T^-)=\Delta Q(V_a,V_b;T^+)+\Delta Q(V_c,V_d;T^-)=\int_{V_a}^{V_b} C_T^{(V)}(V,T^+)dV+\int_{V_c}^{V_d} C_T^{(V)}(V,T^-)dV .$$

This quantity is used by thermodynamics and is related in a special way to the net work done by the body during the Carnot cycle. The net change of the body's internal energy during the Carnot cycle, $\Delta U(V_a,V_b;T^+;V_c,V_d;T^-)$, is equal to zero, because the material of the working body has the special properties noted above.

Special Interest of Calorimetry in Thermodynamics: Relations between Classical Calorimetric Quantities

Relation of Latent Heat with Respect to Volume, and the Equation of State

The quantity $C_T^{(V)}(V,T)$, the latent heat with respect to volume, belongs to classical calorimetry. It accounts for the occurrence of energy transfer by work in a process in which heat is also transferred; the quantity, however, was considered before the relation between heat and work transfers was clarified by the invention of thermodynamics. In the light of thermodynamics, the classical calorimetric quantity is revealed as being tightly linked to the calorimetric material's equation of state $p = p(V,T)$. Provided that the temperature T is measured in the thermodynamic absolute scale, the relation is expressed in the formula

$$C_T^{(V)}(V,T) = T \frac{\partial p}{\partial T}\bigg|_{(V,T)}.$$

Difference of Specific Heats

Advanced thermodynamics provides the relation

$$C_p(p,T) - C_V(V,T) = \left[p(V,T) + \frac{\partial U}{\partial V}\bigg|_{(V,T)} \right] \frac{\partial V}{\partial T}\bigg|_{(p,T)}.$$

From this, further mathematical and thermodynamic reasoning leads to another relation between classical calorimetric quantities. The difference of specific heats is given by

$$C_p(p,T) - C_V(V,T) = \frac{TV \beta_p^2(T,p)}{\kappa_T(T,p)}.$$

Practical Constant-volume Calorimetry (Bomb Calorimetry) for Thermodynamic Studies

Constant-volume calorimetry is calorimetry performed at a constant volume. This involves the use of a constant-volume calorimeter.

No work is performed in constant-volume calorimetry, so the heat measured equals the change in internal energy of the system. The heat capacity at constant volume is assumed to be independent of temperature.

Heat is measured by the principle of calorimetry.

$$q = C_V \Delta T = \Delta U,$$

where

   ΔU is change in internal energy,

   ΔT is change in temperature and

   C_V is the heat capacity at constant volume.

In *constant-volume calorimetry* the pressure is not held constant. If there is a pressure difference between initial and final states, the heat measured needs adjustment to provide the *enthalpy change*. One then has

$$\Delta H = \Delta U + \Delta(PV) = \Delta U + V\Delta P,$$

where

ΔH is change in enthalpy and

V is the unchanging volume of the sample chamber.

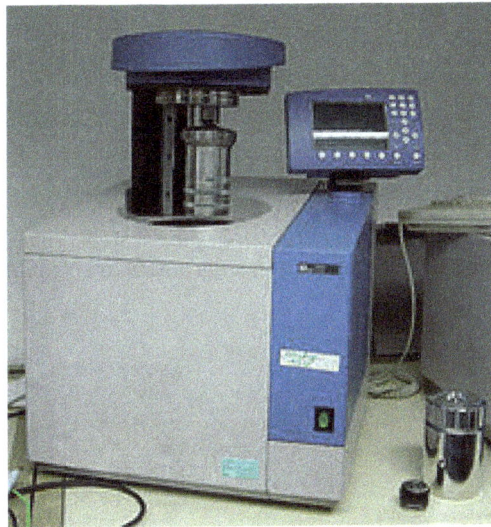

Chromatography

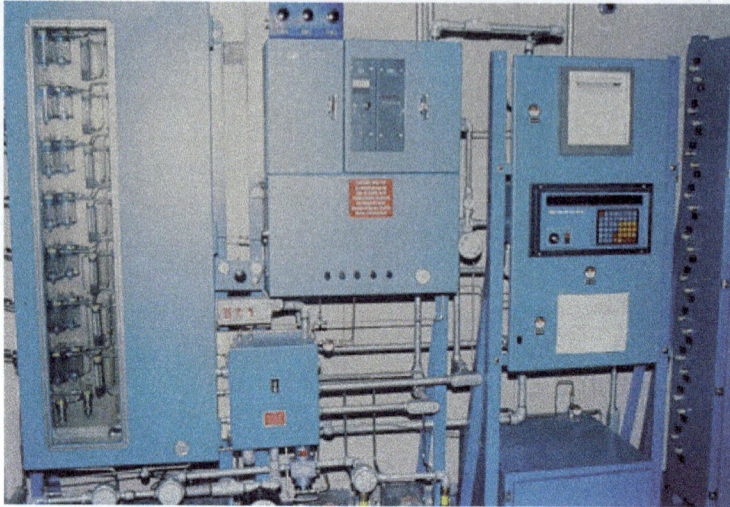

Pictured is a sophisticated gas chromatography system. This instrument records concentrations of acrylonitrile in the air at various points throughout the chemical laboratory.

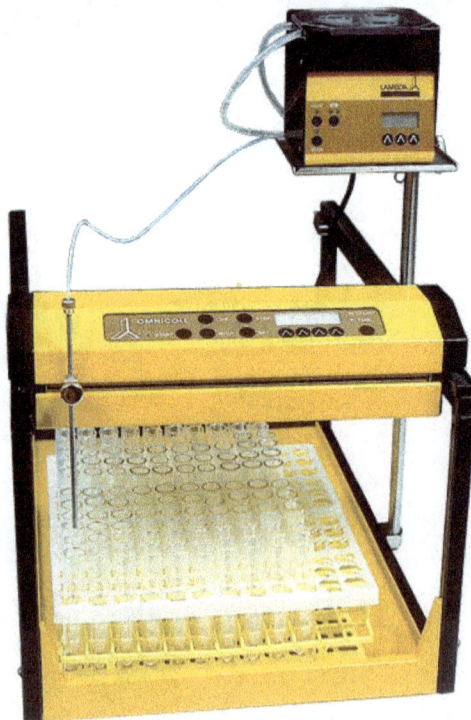

Automated fraction collector and sampler for chromatographic techniques

Chromatography is the collective term for a set of laboratory techniques for the separation of mixtures. The mixture is dissolved in a fluid called the *mobile phase,* which carries it through a structure holding another material called the *stationary phase*. The various constituents of the mixture travel at different speeds, causing them to separate. The separation is based on differential partitioning between the mobile and stationary phases. Subtle differences in a compound's partition coefficient result in differential retention on the stationary phase and thus changing the separation.

Chromatography may be preparative or analytical. The purpose of preparative chromatography is to separate the components of a mixture for more advanced use (and is thus a form of purification). Analytical chromatography is done normally with smaller amounts of material and is for measuring the relative proportions of analytes in a mixture. The two are not mutually exclusive.

History

Thin layer chromatography is used to separate components of a plant extract, illustrating the experiment with plant pigments that gave chromatography its name

Chromatography was first employed in Russia by the Italian-born scientist Mikhail Tsvet in 1900. He continued to work with chromatography in the first decade of the 20th century, primarily for the separation of plant pigments such as chlorophyll, carotenes, and xanthophylls. Since these components have different colors (green, orange, and yellow, respectively) they gave the technique its name. New types of chromatography developed during the 1930s and 1940s made the technique useful for many separation processes.

Chromatography technique developed substantially as a result of the work of Archer John Porter Martin and Richard Laurence Millington Synge during the 1940s and 1950s, for which they won a Nobel prize. They established the principles and basic techniques of partition chromatography, and their work encouraged the rapid development of several chromatographic methods: paper chromatography, gas chromatography, and what would become known as high performance liquid chromatography. Since then, the technology has advanced rapidly. Researchers found that the main principles of Tsvet's chromatography could be applied in many different ways, resulting in the different varieties of chromatography described below. Advances are continually improving the technical performance of chromatography, allowing the separation of increasingly similar molecules.

Chromatography Terms

- The analyte is the substance to be separated during chromatography. It is also normally what is needed from the mixture.

- Analytical chromatography is used to determine the existence and possibly also the concentration of analyte(s) in a sample.

- A bonded phase is a stationary phase that is covalently bonded to the support particles or to the inside wall of the column tubing.

- A chromatogram is the visual output of the chromatograph. In the case of an optimal separation, different peaks or patterns on the chromatogram correspond to different components of the separated mixture.

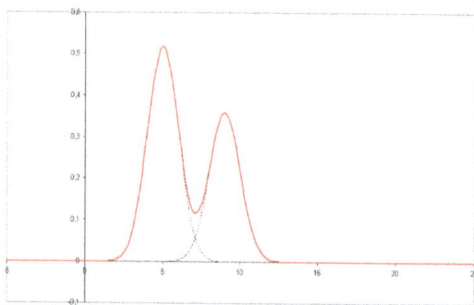

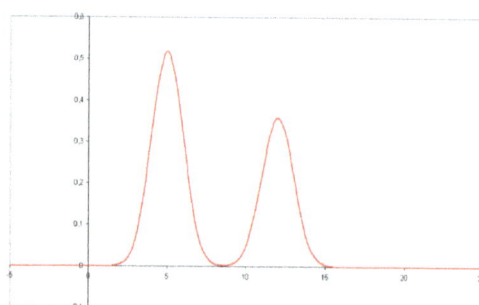

Plotted on the x-axis is the retention time and plotted on the y-axis a signal (for example obtained by a spectrophotometer, mass spectrometer or a variety of other detectors) corresponding to the response created by the analytes exiting the system. In the case of an optimal system the signal is proportional to the concentration of the specific analyte separated.

- A chromatograph is equipment that enables a sophisticated separation, e.g. gas chromatographic or liquid chromatographic separation.

- Chromatography is a physical method of separation that distributes components to separate between two phases, one stationary (stationary phase), the other (the mobile phase) moving in a definite direction.

- The eluate is the mobile phase leaving the column.

- The eluent is the solvent that carries the analyte.

- An eluotropic series is a list of solvents ranked according to their eluting power.

- An immobilized phase is a stationary phase that is immobilized on the support particles, or on the inner wall of the column tubing.

- The mobile phase is the phase that moves in a definite direction. It may be a liquid (LC and Capillary Electrochromatography (CEC)), a gas (GC), or a supercritical fluid (supercritical-fluid chromatography, SFC). The mobile phase consists of the sample being separated/analyzed and the solvent that moves the sample through the column. In the case of HPLC the mobile phase consists of a non-polar solvent(s) such as hexane in normal phase or a polar solvent such as methanol in reverse phase chromatography and the sample being separated. The mobile phase moves through the chromatography column (the stationary phase) where the sample interacts with the stationary phase and is separated.

- Preparative chromatography is used to purify sufficient quantities of a substance for further use, rather than analysis.

- The retention time is the characteristic time it takes for a particular analyte to pass through the system (from the column inlet to the detector) under set conditions.

- The sample is the matter analyzed in chromatography. It may consist of a single component or it may be a mixture of components. When the sample is treated in the course of an analysis, the phase or the phases containing the analytes of interest is/are referred to as the sample whereas everything out of interest separated from the sample before or in the course of the analysis is referred to as waste.

- The solute refers to the sample components in partition chromatography.

- The solvent refers to any substance capable of solubilizing another substance, and especially the liquid mobile phase in liquid chromatography.

- The stationary phase is the substance fixed in place for the chromatography procedure. Examples include the silica layer in thin layer chromatography

- The detector refers to the instrument used for qualitative and quantitative detection of analytes after separation.

Chromatography is based on the concept of partition coefficient. Any solute partitions between two immiscible solvents. When we make one solvent immobile (by adsorption on a solid support matrix) and another mobile it results in most common applications of chromatography. If the matrix support, or stationary phase, is polar (e.g. paper, silica etc.) it is forward phase chromatography, and if it is non-polar (C-18) it is reverse phase.

Techniques by Chromatographic Bed Shape

Column Chromatography

Column chromatography is a separation technique in which the stationary bed is within a tube. The particles of the solid stationary phase or the support coated with a liquid stationary phase may fill the whole inside volume of the tube (packed column) or be concentrated on or along the inside tube wall leaving an open, unrestricted path for the mobile phase in the middle part of the tube (open tubular column). Differences in rates of movement through the medium are calculated to different retention times of the sample.

In 1978, W. Clark Still introduced a modified version of column chromatography called flash column chromatography (flash). The technique is very similar to the traditional column chromatography, except for that the solvent is driven through the column by applying positive pressure. This allowed most separations to be performed in less than 20 minutes, with improved separations compared to the old method. Modern flash chromatography systems are sold as pre-packed plastic cartridges, and the solvent is pumped through the cartridge. Systems may also be linked with detectors and fraction collectors providing automation. The introduction of gradient pumps resulted in quicker separations and less solvent usage.

In expanded bed adsorption, a fluidized bed is used, rather than a solid phase made by a packed bed. This allows omission of initial clearing steps such as centrifugation and filtration, for culture broths or slurries of broken cells.

Phosphocellulose chromatography utilizes the binding affinity of many DNA-binding proteins for phosphocellulose. The stronger a protein's interaction with DNA, the higher the salt concentration needed to elute that protein.

Planar Chromatography

Planar chromatography is a separation technique in which the stationary phase is present as or on a plane. The plane can be a paper, serving as such or impregnated by a substance as the stationary bed (paper chromatography) or a layer of solid particles spread on a support such as a glass plate (thin layer chromatography). Different compounds in the sample mixture travel different distances according to how strongly they interact with the stationary phase as compared to the mobile phase. The specific Retention factor (R_f) of each chemical can be used to aid in the identification of an unknown substance.

Paper Chromatography

Paper chromatography is a technique that involves placing a small dot or line of sample solution onto a strip of *chromatography paper*. The paper is placed in a container with a shallow layer of solvent and sealed. As the solvent rises through the paper, it meets the sample mixture, which starts to travel up the paper with the solvent. This paper is made of cellulose, a polar substance, and the compounds within the mixture travel farther if they are non-polar. More polar substances bond with the cellulose paper more quickly, and therefore do not travel as far.

Thin Layer Chromatography (TLC)

Thin layer chromatography (TLC) is a widely employed laboratory technique and is similar to paper chromatography. However, instead of using a stationary phase of paper, it involves a stationary phase of a thin layer of adsorbent like silica gel, alumina, or cellulose on a flat, inert substrate. Compared to paper, it has the advantage of faster runs, better separations, and the choice between different adsorbents. For even better resolution and to allow for quantification, high-performance TLC can be used. An older popular use had been to differentiate chromosomes by observing distance in gel (separation of was a separate step).

Displacement Chromatography

The basic principle of displacement chromatography is: A molecule with a high affinity for the chromatography matrix (the displacer) competes effectively for binding sites, and thus displace all molecules with lesser affinities. There are distinct differences between displacement and elution chromatography. In elution mode, substances typically emerge from a column in narrow, Gaussian peaks. Wide separation of peaks, preferably to baseline, is desired for maximum purification. The speed at which any component of a mixture travels down the column in elution mode depends on many factors. But for two substances to travel at different speeds, and thereby be resolved, there must be substantial differences in some interaction between the biomolecules and the chromatography matrix. Operating parameters are adjusted to maximize the effect of this difference. In many cases, baseline separation of the peaks can be achieved only with gradient elution and low column loadings. Thus, two drawbacks to elution mode chromatography, especially at the preparative scale, are operational complexity, due to gradient solvent pumping, and low throughput, due to low column loadings. Displacement chromatography has advantages over elution chromatography in that components are resolved into consecutive zones of pure substances rather than "peaks". Because the process takes advantage of the nonlinearity of the isotherms, a larger column feed can be separated on a given column with the purified components recovered at significantly higher concentrations.

Techniques by Physical state of Mobile Phase

Gas Chromatography

Gas chromatography (GC), also sometimes known as gas-liquid chromatography,

(GLC), is a separation technique in which the mobile phase is a gas. Gas chromatographic separation is always carried out in a column, which is typically "packed" or "capillary". Packed columns are the routine work horses of gas chromatography, being cheaper and easier to use and often giving adequate performance. Capillary columns generally give far superior resolution and although more expensive are becoming widely used, especially for complex mixtures. Both types of column are made from non-adsorbent and chemically inert materials. Stainless steel and glass are the usual materials for packed columns and quartz or fused silica for capillary columns.

Gas chromatography is based on a partition equilibrium of analyte between a solid or viscous liquid stationary phase (often a liquid silicone-based material) and a mobile gas (most often helium). The stationary phase is adhered to the inside of a small-diameter (commonly 0.53 – 0.18mm inside diameter) glass or fused-silica tube (a capillary column) or a solid matrix inside a larger metal tube (a packed column). It is widely used in analytical chemistry; though the high temperatures used in GC make it unsuitable for high molecular weight biopolymers or proteins (heat denatures them), frequently encountered in biochemistry, it is well suited for use in the petrochemical, environmental monitoring and remediation, and industrial chemical fields. It is also used extensively in chemistry research.

Liquid Chromatography

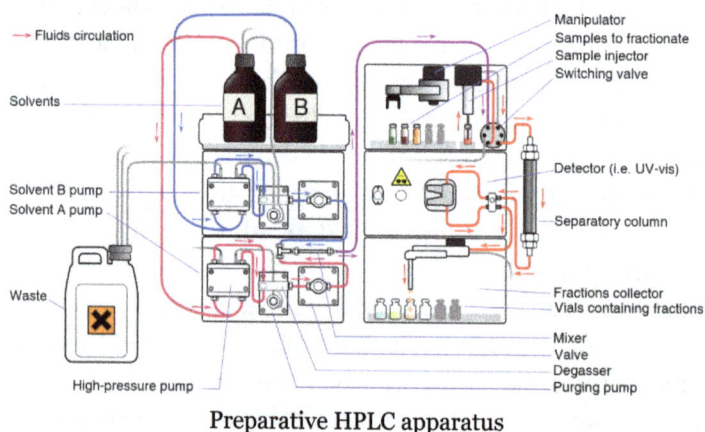

Preparative HPLC apparatus

Liquid chromatography (LC) is a separation technique in which the mobile phase is a liquid. It can be carried out either in a column or a plane. Present day liquid chromatography that generally utilizes very small packing particles and a relatively high pressure is referred to as high performance liquid chromatography (HPLC).

In HPLC the sample is forced by a liquid at high pressure (the mobile phase) through a column that is packed with a stationary phase composed of irregularly or spherically shaped particles, a porous monolithic layer, or a porous membrane. HPLC is historically divided into two different sub-classes based on the polarity of the mobile and stationary phases. Methods in which the stationary phase is more polar than the mobile phase (e.g., toluene as the mobile phase, silica as the stationary phase) are termed normal phase liquid chromatography (NPLC) and the opposite (e.g., water-methanol mixture as the mobile phase and C18 = octadecylsilyl as the stationary phase) is termed reversed phase liquid chromatography (RPLC).

Specific techniques under this broad heading are listed below.

Affinity Chromatography

Affinity chromatography is based on selective non-covalent interaction between an analyte and specific molecules. It is very specific, but not very robust. It is often used in biochemistry in the purification of proteins bound to tags. These fusion proteins are labeled with compounds such as His-tags, biotin or antigens, which bind to the stationary phase specifically. After purification, some of these tags are usually removed and the pure protein is obtained.

Affinity chromatography often utilizes a biomolecule's affinity for a metal (Zn, Cu, Fe, etc.). Columns are often manually prepared. Traditional affinity columns are used as a preparative step to flush out unwanted biomolecules.

However, HPLC techniques exist that do utilize affinity chromatography properties. Immobilized Metal Affinity Chromatography (IMAC) is useful to separate aforementioned molecules based on the relative affinity for the metal (I.e. Dionex IMAC). Often these columns can be loaded with different metals to create a column with a targeted affinity.

Supercritical Fluid Chromatography

Supercritical fluid chromatography is a separation technique in which the mobile phase is a fluid above and relatively close to its critical temperature and pressure.

Techniques by Separation Mechanism

Ion Exchange Chromatography

Ion exchange chromatography (usually referred to as ion chromatography) uses an ion exchange mechanism to separate analytes based on their respective charges. It is usually performed in columns but can also be useful in planar mode. Ion exchange chromatography uses a charged stationary phase to separate charged compounds including anions, cations, amino acids, peptides, and proteins. In conventional methods the stationary phase is an ion exchange resin that carries charged functional groups that interact with oppositely charged groups of the compound to retain. Ion exchange chromatography is commonly used to purify proteins using FPLC.

Size-exclusion chromatography

Size-exclusion chromatography (SEC) is also known as gel permeation chromatography (GPC) or gel filtration chromatography and separates molecules according to their size (or more accurately according to their hydrodynamic diameter or hydrodynamic volume). Smaller molecules are able to enter the pores of the media and, therefore, molecules are trapped and removed from the flow of the mobile phase. The average residence time in the pores depends upon the effective size of the analyte molecules. However, molecules that are larger than the average pore size of the packing are excluded and thus suffer essentially no retention; such species are the first to be eluted. It is generally a low-resolution chromatography technique and thus it is often reserved for the final, "polishing" step of a purification. It is also useful for determining the tertiary structure and quaternary structure of purified proteins, especially since it can be carried out under native solution conditions.

Expanded Bed Adsorption Chromatographic Separation

An expanded bed chromatographic adsorption (EBA) column for a biochemical separation process comprises a pressure equalization liquid distributor having a self-cleaning function below a porous blocking sieve plate at the bottom of the expanded bed, an upper part nozzle assembly having a backflush cleaning function at the top of the expanded bed, a better distribution of the feedstock liquor added into the expanded bed ensuring that the fluid passed through the expanded bed layer displays a state of piston flow. The expanded bed layer displays a state of piston flow. The expanded bed chromatographic separation column has advantages of increasing the separation efficiency of the expanded bed.

Expanded-bed adsorption (EBA) chromatography is a convenient and effective technique for the capture of proteins directly from unclarified crude sample. In EBA chromatography, the settled bed is first expanded by upward flow of equilibration buffer. The crude feed, a mixture of soluble proteins, contaminants, cells, and cell debris, is then passed upward through the expanded bed. Target proteins are captured on the adsorbent, while particulates and contaminants pass through. A change to elution buffer while maintaining upward flow results in desorption of the target protein in expanded-bed mode. Alternatively, if the flow is reversed, the adsorbed particles will quickly settle and the proteins can be desorbed by an elution buffer. The mode used for elution (expanded-bed versus settled-bed) depends on the characteristics of the feed. After elution, the adsorbent is cleaned with a predefined cleaning-in-place (CIP) solution, with cleaning followed by either column regeneration (for further use) or storage.

Special Techniques

Reversed-phase Chromatography

Reversed-phase chromatography (RPC) is any liquid chromatography procedure in which the mobile phase is significantly more polar than the stationary phase. It is so named because in normal-phase liquid chromatography, the mobile phase is significantly less polar than the stationary phase. Hydrophobic molecules in the mobile phase tend to adsorb to the relatively hydrophobic stationary phase. Hydrophilic molecules in the mobile phase will tend to elute first. Separating columns typically comprise a C8 or C18 carbon-chain bonded to a silica particle substrate.

Hydrophobic Interaction Chromatography

Hydrophobic interactions between proteins and the chromatographic matrix can be exploited to purify proteins. In hydrophobic interaction chromatography the matrix material is lightly substituted with hydrophobic groups. These groups can range from methyl, ethyl, propyl, octyl, or phenyl groups.[] At high salt concentrations, non-polar sidechains on the surface on proteins "interact" with the hydrophobic groups; that is, both types of groups are excluded by the polar solvent (hydrophobic effects are augmented by increased ionic strength). Thus, the sample is applied to the column in a buffer which is highly polar. The eluant is typically an aqueous buffer with decreasing salt concentrations, increasing concentrations of detergent (which disrupts hydrophobic interactions), or changes in pH.

In general, Hydrophobic Interaction Chromatography (HIC) is advantageous if the sample is sensitive to pH change or harsh solvents typically used in other types of chromatography but not high salt concentrations. Commonly, it is the amount of salt in the buffer which is varied. In 2012, Müller and Franzreb described the effects of temperature on HIC using Bovine Serum Albumin (BSA) with four different types of hydrophobic resin. The study altered temperature as to effect the binding affinity of BSA onto the matrix. It was concluded that cycling temperature from 50 degrees to 10 degrees would not be adequate to effectively wash all BSA from the matrix but could be very effective if the column would only be used a few times. Using temperature to effect change allows labs to cut costs on buying salt and saves money.

Two-dimensional chromatograph GCxGC-TOFMS at Chemical Faculty of GUTGdańsk, Poland, 2016

If high salt concentrations along with temperature fluctuations want to be avoided you can use a more hydrophobic to compete with your sample to elute it. [source] This so-called

Countercurrent chromatography (CCC) is a type of liquid-liquid chromatography, where both the stationary and mobile phases are liquids. The operating principle of CCC equipment requires a column consisting of an open tube coiled around a bobbin. The bobbin is rotated in a double-axis gyratory motion (a cardioid), which causes a variable gravity (G) field to act on the column during each rotation. This motion causes the column to see one partitioning step per revolution and components of the sample separate in the column due to their partitioning coefficient between the two immiscible liquid phases used. There are many types of CCC available today. These include HSCCC (High Speed CCC) and HPCCC (High Performance CCC). HPCCC is the latest and best performing version of the instrumentation available currently.

Chiral Chromatography

Chiral chromatography involves the separation of stereoisomers. In the case of enantiomers, these have no chemical or physical differences apart from being three-dimensional mirror images. Conventional chromatography or other separation processes are incapable of separating them. To enable chiral separations to take place, either the mobile phase or the stationary phase must themselves be made chiral, giving differing affinities between the analytes. Chiral chromatography HPLC columns (with a chiral stationary phase) in both normal and reversed phase are commercially available.

Circular Dichroism

Circular dichroism (CD) is dichroism involving circularly polarized light, i.e., the differential absorption of left- and right-handed light. Left-hand circular (LHC) and right-hand circular (RHC) polarized light represent two possible spin angular momentum states for a photon, and so circular dichroism is also referred to as dichroism for spin angular momentum. This phenomenon was discovered by Jean-Baptiste Biot, Augustin Fresnel, and Aimé Cotton in the first half of the 19th century. It is exhibited in the absorption bands of optically activechiral molecules. CD spectroscopy has a wide range of applications in many different fields. Most notably, UV CD is used to investigate the secondary structure of proteins. UV/Vis CD is used to investigate charge-transfer transitions.Near-infrared CD is used to investigate geometric and electronic structure by probing metald→d transitions.Vibrational circular dichroism, which uses light from the infrared energy region, is used for structural studies of small organic molecules, and most recently proteins and DNA.

Physical Principles

Circular Polarization of Light

Electromagnetic radiation consists of an electric (E) and magnetic (B) field that oscillate perpendicular to one another and to the propagating direction, a transverse wave. While linearly polarized light occurs when the electric field vector oscillates only in one plane, circularly polarized light occurs when the direction of the electric field vector rotates about its propagation direction while the vector retains constant magnitude. At a single point in space, the circularly polarized-vector will trace out a circle over one period of the wave frequency, hence the name. The two diagrams below show the electric vectors of linearly and circularly polarized light, at one moment of time, for a range of positions; the plot of the circularly polarized electric vector forms a helix along the direction of propagation (k). For left circularly polarized light (LCP) with propagation towards the observer, the electric vector rotates counterclockwise. For right circularly polarized light (RCP), the electric vector rotates clockwise.

Linearly polarized

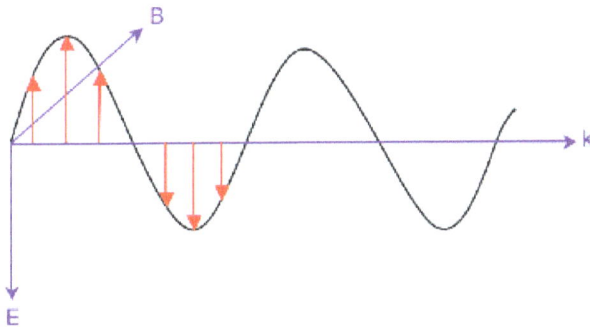

Circularly polarized

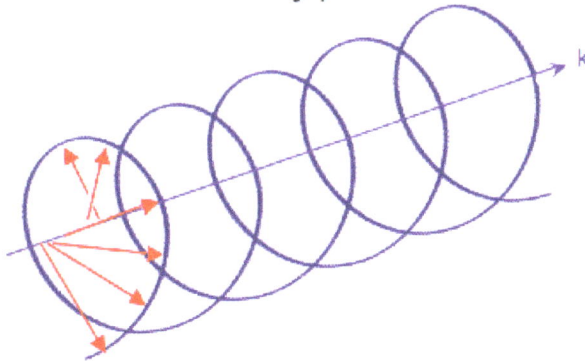

Interaction of Circularly Polarized Light with Matter

When circularly polarized light passes through an absorbing optically active medium, the speeds between right and left polarizations differ ($c_L \neq c_R$) as well as their wavelength ($\lambda_L \neq \lambda_R$) and the extent to which they are absorbed ($\varepsilon_L \neq \varepsilon_R$). *Circular dichroism* is the difference $\Delta\varepsilon \equiv \varepsilon_L - \varepsilon_R$. The electric field of a light beam causes a linear displacement of charge when interacting with a molecule (electric dipole), whereas its magnetic field causes a circulation of charge (magnetic dipole). These two motions combined cause an excitation of an electron in a helical motion, which includes translation and rotation and their associated operators. The experimentally determined relationship between the rotational strength (R) of a sample and the $\Delta \varepsilon$ is given by

$$R_{exp} = \frac{3hc10^3 \ln(10)}{32\pi^3 N_A} \int \frac{\Delta \epsilon}{v} dv$$

The rotational strength has also been determined theoretically,

$$R_{theo} = \frac{1}{2mc} Im \int \Psi_g \hat{M}_{(elec.dipole)} \Psi_e d\tau \cdot \int \Psi_g \hat{M}_{(mag.dipole)} \Psi_e d\tau$$

We see from these two equations that in order to have non-zero $\Delta \epsilon$, the electric and magnetic dipole moment operators ($\hat{M}_{(elec.dipole)}$ and $\hat{M}_{(mag.dipole)}$) must transform as the same irreducible representation. C_n and D_n are the only point groups where this can occur, making only chiral molecules CD active.

Simply put, since circularly polarized light itself is "chiral", it interacts differently with chiral molecules. That is, the two types of circularly polarized light are absorbed to different extents. In a CD experiment, equal amounts of left and right circularly polarized light of a selected wavelength are alternately radiated into a (chiral) sample. One of the two polarizations is absorbed more than the other one, and this wavelength-dependent difference of absorption is measured, yielding the CD spectrum of the sample. Due to the interaction with the molecule, the electric field vector of the light traces out an elliptical path after passing through the sample.

It is important that the chirality of the molecule can be conformational rather than structural. That is, for instance, a protein molecule with a helical secondary structure can have a CD that changes with changes in the conformation.

Delta Absorbance

By definition,

$$\Delta A = A_L - A_R$$

where ΔA (Delta Absorbance) is the difference between absorbance of left circularly polarized (LCP) and right circularly polarized (RCP) light (this is what is usually measured). ΔA is a function of wavelength, so for a measurement to be meaningful the wavelength at which it was performed must be known.

Molar Circular Dichroism

It can also be expressed, by applying Beer's law, as:

$$\Delta A = (\epsilon_L - \epsilon_R)Cl$$

where

ϵ_L and ϵ_R are the molar extinction coefficients for LCP and RCP light,

C is the molar concentration,

l is the path length in centimeters (cm).

Then

$$\Delta\epsilon = \epsilon_L - \epsilon_R$$

is the molar circular dichroism. This intrinsic property is what is usually meant by the circular dichroism of the substance. Since Äò is a function of wavelength, a molar circular dichroism value ($\Delta\epsilon$) must specify the wavelength at which it is valid.

Extrinsic Effects on Circular Dichroism

In many practical applications of circular dichroism (CD), as discussed below, the measured CD is not simply an intrinsic property of the molecule, but rather depends on the molecular conformation. In such a case the CD may also be a function of temperature, concentration, and the chemical environment, including solvents. In this case the reported CD value must also specify these other relevant factors in order to be meaningful.

Molar Ellipticity

Although ΔA is usually measured, for historical reasons most measurements are reported in degrees of ellipticity. Molar ellipticity is circular dichroism corrected for concentration. Molar circular dichroism and molar ellipticity, [θ], are readily interconverted by the equation:

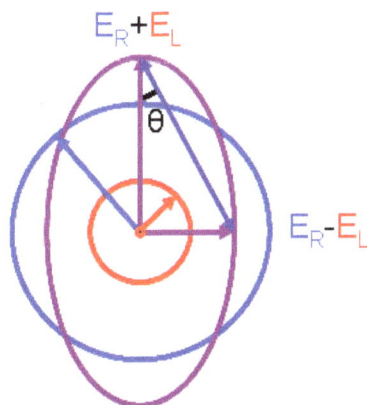

Elliptical polarized light (violet) is composed of unequal
contributions of right (blue) and left (red) circular polarized light.

$$[\theta] = 3298.2\Delta\varepsilon.$$

This relationship is derived by defining the ellipticity of the polarization as:

$$\tan\theta = \frac{E_R - E_L}{E_R + E_L}$$

where

E_R and E_L are the magnitudes of the electric fieldvectors of the right-circularly and left-circularly polarized light, respectively.

When E_R equals E_L (when there is no difference in the absorbance of right- and left-circular polarized light), θ is 0° and the light is linearly polarized. When either E_R or E_L is equal to zero (when there is complete absorbance of the circular polarized light in one direction), θ is 45° and the light is circularly polarized.

Generally, the circular dichroism effect is small, so $\tan\theta$ is small and can be approximated as θ in radians. Since the intensity or irradiance, I, of light is proportional to the square of the electric-field vector, the ellipticity becomes:

$$\theta(\text{radians}) = \frac{(I_R^{1/2} - I_L^{1/2})}{(I_R^{1/2} + I_L^{1/2})}$$

Then by substituting for I using Beer's law in natural logarithm form:

$$I = I_0 e^{-A\ln 10}$$

The ellipticity can now be written as:

$$\theta(\text{radians}) = \frac{(e^{\frac{-A_R}{2}\ln 10} - e^{\frac{-A_L}{2}\ln 10})}{(e^{\frac{-A_R}{2}\ln 10} + e^{\frac{-A_L}{2}\ln 10})} = \frac{e^{\Delta A \frac{\ln 10}{2}} - 1}{e^{\Delta A \frac{\ln 10}{2}} + 1}$$

Since $\Delta A << 1$, this expression can be approximated by expanding the exponentials in a Taylor series to first-order and then discarding terms of ΔA in comparison with unity and converting from radians to degrees:

$$\theta(\text{degrees}) = \Delta A \left(\frac{\ln 10}{4}\right)\left(\frac{180}{\pi}\right)$$

The linear dependence of solute concentration and pathlength is removed by defining molar ellipticity as,

$$[\theta] = \frac{100\theta}{Cl}$$

Then combining the last two expression with Beer's law, molar ellipticity becomes:

$$[\theta] = 100\Delta\varepsilon \left(\frac{\ln 10}{4} \right) \left(\frac{180}{\pi} \right) = 3298.2\Delta\varepsilon$$

The units of molar ellipticity are historically (deg·cm²/dmol). To calculate molar ellipticity, the sample concentration (g/L), cell pathlength (cm), and the molecular weight (g/mol) must be known.

If the sample is a protein, the mean residue weight (average molecular weight of the amino acid residues it contains) is often used in place of the molecular weight, essentially treating the protein as a solution of amino acids. Using mean residue ellipticity facilitates comparing the CD of proteins of different molecular weight; use of this normalized CD is important in studies of protein structure.

Mean Residue Ellipticity

Methods for estimating secondary structure in polymers, proteins and polypeptides in particular, often require that the measured molar ellipticity spectrum be converted to a normalized value, specifically a value independent of the polymer length. Mean residue ellipticity is used for this purpose; it is simply the measured molar ellipticity of the molecule divided by the number of monomer units (residues) in the molecule.

Application to Biological Molecules

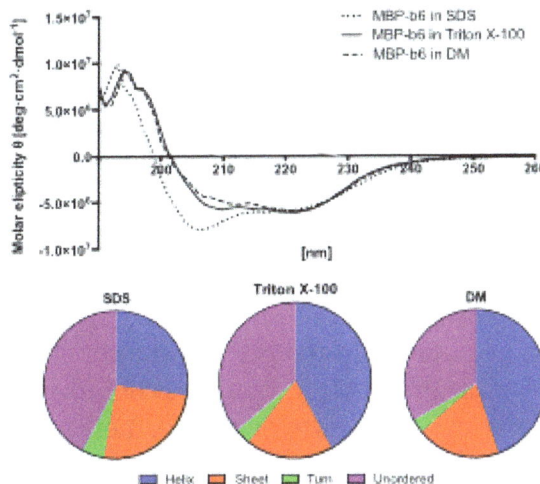

Upper panel: Circular dichroism spectroscopy in the ultraviolet wavelength region (UV-CD) of MBP-cytochrome b_6 fusion protein in different detergent solutions. It shows that the protein in DM, as well as in Triton X-100 solution, recovered its structure. However the spectra obtained from SDS solution shows decreased ellipticity in the range between 200–210 nm, which indicates incomplete secondary structure recovery. Lower panel: The content of secondary structures predicted from the CD spectra using the CDSSTR algorithm. The protein in SDS solution shows increased content of unordered structures and decreased helices content.

In general, this phenomenon will be exhibited in absorption bands of any optically active molecule. As a consequence, circular dichroism is exhibited by biological molecules, because of their dextrorotary and levorotary components. Even more important is that a secondary structure will also impart a distinct CD to its respective molecules. Therefore, the alpha helix of proteins and the double helix of nucleic acids have CD spectral signatures representative of their structures. The capacity of CD to give a representative structural signature makes it a powerful tool in modern biochemistry with applications that can be found in virtually every field of study.

CD is closely related to the optical rotatory dispersion (ORD) technique, and is generally considered to be more advanced. CD is measured in or near the absorption bands of the molecule of interest, while ORD can be measured far from these bands. CD's advantage is apparent in the data analysis. Structural elements are more clearly distinguished since their recorded bands do not overlap extensively at particular wavelengths as they do in ORD. In principle these two spectral measurements can be interconverted through an integral transform (Kramers–Kronig relation), if all the absorptions are included in the measurements.

The far-UV (ultraviolet) CD spectrum of proteins can reveal important characteristics of their secondary structure. CD spectra can be readily used to estimate the fraction of a molecule that is in the alpha-helix conformation, the beta-sheet conformation, the beta-turn conformation, or some other (e.g. random coil) conformation. These fractional assignments place important constraints on the possible secondary conformations that the protein can be in. CD cannot, in general, say where the alpha helices that are detected are located within the molecule or even completely predict how many there are. Despite this, CD is a valuable tool, especially for showing changes in conformation. It can, for instance, be used to study how the secondary structure of a molecule changes as a function of temperature or of the concentration of denaturing agents, e.g. Guanidinium chloride or urea. In this way it can reveal important thermodynamic information about the molecule (such as the enthalpy and Gibbs free energy of denaturation) that cannot otherwise be easily obtained. Anyone attempting to study a protein will find CD a valuable tool for verifying that the protein is in its native conformation before undertaking extensive and/or expensive experiments with it. Also, there are a number of other uses for CD spectroscopy in protein chemistry not related to alpha-helix fraction estimation.

The near-UV CD spectrum (>250 nm) of proteins provides information on the tertiary structure. The signals obtained in the 250–300 nm region are due to the absorption, dipole orientation and the nature of the surrounding environment of the phenylalanine, tyrosine, cysteine (or S-S disulfide bridges) and tryptophan amino acids. Unlike in far-UV CD, the near-UV CD spectrum cannot be assigned to any particular 3D structure. Rather, near-UV CD spectra provide structural information on the nature of the prosthetic groups in proteins, e.g., the heme groups in hemoglobin and cytochrome c.

Visible CD spectroscopy is a very powerful technique to study metal–protein interactions and can resolve individual d–d electronic transitions as separate bands. CD spectra in the visible light region are only produced when a metal ion is in a chiral environment, thus, free metal ions in solution are not detected. This has the advantage of only observing the protein-bound metal, so pH dependence and stoichiometries are readily obtained. Optical activity in transition metal ion complexes have been attributed to configurational, conformational and the vicinal effects. Klewpatinond and Viles (2007) have produced a set of empirical rules for predicting the appearance of visible CD spectra for Cu^{2+} and Ni^{2+} square-planar complexes involving histidine and main-chain coordination.

CD gives less specific structural information than X-ray crystallography and protein NMR spectroscopy, for example, which both give atomic resolution data. However, CD spectroscopy is a quick method that does not require large amounts of proteins or extensive data processing. Thus CD can be used to survey a large number of solvent conditions, varying temperature, pH, salinity, and the presence of various cofactors.

CD spectroscopy is usually used to study proteins in solution, and thus it complements methods that study the solid state. This is also a limitation, in that many proteins are embedded in membranes in their native state, and solutions containing membrane structures are often strongly scattering. CD is sometimes measured in thin films.

Experimental Limitations

CD has also been studied in carbohydrates, but with limited success due to the experimental difficulties associated with measurement of CD spectra in the vacuum ultraviolet (VUV) region of the spectrum (100–200 nm), where the corresponding CD bands of unsubstituted carbohydrates lie. Substituted carbohydrates with bands above the VUV region have been successfully measured.

Measurement of CD is also complicated by the fact that typical aqueous buffer systems often absorb in the range where structural features exhibit differential absorption of circularly polarized light. Phosphate, sulfate, carbonate, and acetate buffers are generally incompatible with CD unless made extremely dilute e.g. in the 10–50 mM range. The TRIS buffer system should be completely avoided when performing far-UV CD. Borate and Onium compounds are often used to establish the appropriate pH range for CD experiments. Some experimenters have substituted fluoride for chloride ion because fluoride absorbs less in the far UV, and some have worked in pure water. Another, almost universal, technique is to minimize solvent absorption by using shorter path length cells when working in the far UV, 0.1 mm path lengths are not uncommon in this work.

In addition to measuring in aqueous systems, CD, particularly far-UV CD, can be measured in organic solvents e.g. ethanol, methanol, trifluoroethanol (TFE). The latter has the advantage to induce structure formation of proteins, inducing beta-sheets in some and alpha helices in others, which they would not show under normal aqueous conditions. Most common organic solvents such as acetonitrile, THF, chloroform, dichloromethane are however, incompatible with far-UV CD.

It may be of interest to note that the protein CD spectra used in secondary structure estimation are related to the π to π* orbital absorptions of the amide bonds linking the amino acids. These absorption bands lie partly in the *so-called*vacuum ultraviolet (wavelengths less than about 200 nm). The wavelength region of interest is actually inaccessible in air because of the strong absorption of light by oxygen at these wavelengths. In practice these spectra are measured not in vacuum but in an oxygen-free instrument (filled with pure nitrogen gas).

Once oxygen has been eliminated, perhaps the second most important technical factor in working below 200 nm is to design the rest of the optical system to have low losses in this region. Critical in this regard is the use of aluminized mirrors whose coatings have been optimized for low loss in this region of the spectrum.

The usual light source in these instruments is a high pressure, short-arc xenon lamp. Ordinary xenon arc lamps are unsuitable for use in the low UV. Instead, specially constructed lamps with envelopes made from high-purity synthetic fused silica must be used.

Light from synchrotron sources has a much higher flux at short wavelengths, and has been used to record CD down to 160 nm. In 2010 the CD spectrophotometer at the electron storage ring facility ISA at the University of Aarhus in Denmark was used to record solid state CD spectra down to 120 nm. At the quantum mechanical level, the feature density of circular dichroism and optical rotation are identical. Optical rotary dispersion and circular dichroism share the same quantum information content.

Computational Chemistry

Computational chemistry is a branch of chemistry that uses computer simulation to assist in solving chemical problems. It uses methods of theoretical chemistry, incorporated into efficient computer programs, to calculate the structures and properties of molecules and solids. It is necessary because, apart from relatively recent results concerning the hydrogen molecular ion, the quantum many-body problem cannot be solved analytically, much less in closed form. While computational results normally complement the information obtained by chemical experiments, it can in some cases predict hitherto unobserved chemical phenomena. It is widely used in the design of new drugs and materials.

Examples of such properties are structure (i.e., the expected positions of the constituent atoms), absolute and relative (interaction) energies, electroniccharge density distributions, dipoles and higher multipole moments, vibrational frequencies, reactivity, or other spectroscopic quantities, and cross sections for collision with other particles.

The methods used cover both static and dynamic situations. In all cases, the computer time and other resources (such as memory and disk space) increase rapidly with the size of the system being studied. That system can be one molecule, a group of molecules, or a solid. Computational chemistry methods range from very approximate to highly accurate; the latter are usually feasible for small systems only. *Ab initio* methods are based entirely on quantum mechanics and basic physical constants. Other methods are called empirical or semi-empirical because they use additional empirical parameters.

Both *ab initio* and semi-empirical approaches involve approximations. These range from simplified forms of the first-principles equations that are easier or faster to solve, to approximations limiting the size of the system (for example, periodic boundary conditions), to fundamental approximations to the underlying equations that are required to achieve any solution to them at all. For example, most *ab initio* calculations make the Born–Oppenheimer approximation, which greatly simplifies the underlying Schrödinger equation by assuming that the nuclei remain in place during the calculation. In principle, *ab initio* methods eventually converge to the exact solution of the underlying equations as the number of approximations is reduced. In practice, however, it is impossible to eliminate all approximations, and residual error inevitably remains. The goal of computational chemistry is to minimize this residual error while keeping the calculations tractable.

In some cases, the details of electronic structure are less important than the long-time phase space behavior of molecules. This is the case in conformational studies of proteins and protein-ligand binding thermodynamics. Classical approximations to the potential energy surface are used, as they are computationally less intensive than electronic calculations, to enable longer simulations of molecular dynamics. Furthermore, cheminformatics uses even more empirical (and computationally cheaper) methods like machine learning based on physicochemical properties. One typical problem in cheminformatics is to predict the binding affinity of drug molecules to a given target.

History

Building on the founding discoveries and theories in the history of quantum mechanics, the first theoretical calculations in chemistry were those of Walter Heitler and Fritz

London in 1927. The books that were influential in the early development of computational quantum chemistry include Linus Pauling and E. Bright Wilson's 1935 *Introduction to Quantum Mechanics – with Applications to Chemistry*, Eyring, Walter and Kimball's 1944 *Quantum Chemistry*, Heitler's 1945 *Elementary Wave Mechanics – with Applications to Quantum Chemistry*, and later Coulson's 1952 textbook *Valence*, each of which served as primary references for chemists in the decades to follow.

With the development of efficient computer technology in the 1940s, the solutions of elaborate wave equations for complex atomic systems began to be a realizable objective. In the early 1950s, the first semi-empirical atomic orbital calculations were performed. Theoretical chemists became extensive users of the early digital computers. A very detailed account of such use in the United Kingdom is given by Smith and Sutcliffe. The first *ab initio*Hartree–Fock method calculations on diatomic molecules were performed in 1956 at MIT, using a basis set of Slater orbitals. For diatomic molecules, a systematic study using a minimum basis set and the first calculation with a larger basis set were published by Ransil and Nesbet respectively in 1960. The first polyatomic calculations using Gaussian orbitals were performed in the late 1950s. The first configuration interaction calculations were performed in Cambridge on the EDSAC computer in the 1950s using Gaussian orbitals by Boys and coworkers. By 1971, when a bibliography of *ab initio* calculations was published, the largest molecules included were naphthalene and azulene. Abstracts of many earlier developments in *ab initio* theory have been published by Schaefer.

In 1964, Hückel method calculations (using a simple linear combination of atomic orbitals (LCAO) method to determine electron energies of molecular orbitals of π electrons in conjugated hydrocarbon systems) of molecules, ranging in complexity from butadiene and benzene to ovalene, were generated on computers at Berkeley and Oxford. These empirical methods were replaced in the 1960s by semi-empirical methods such as CNDO.

In the early 1970s, efficient *ab initio* computer programs such as ATMOL, Gaussian, IBMOL, and POLYAYTOM, began to be used to speed *ab initio* calculations of molecular orbitals. Of these four programs, only Gaussian, now vastly expanded, is still in use, but many other programs are now in use. At the same time, the methods of molecular mechanics, such as MM2 force field, were developed, primarily by Norman Allinger.

One of the first mentions of the term *computational chemistry* can be found in the 1970 book *Computers and Their Role in the Physical Sciences* by Sidney Fernbach and Abraham Haskell Taub, where they state "It seems, therefore, that 'computational chemistry' can finally be more and more of a reality." During the 1970s, widely different methods began to be seen as part of a new emerging discipline of *computational chemistry*. The *Journal of Computational Chemistry* was first published in 1980.

Computational chemistry has featured in several Nobel Prize awards, most notably in 1998 and 2013. Walter Kohn, "for his development of the density-functional theory", and John Pople, "for his development of computational methods in quantum chemistry", received the 1998 Nobel Prize in Chemistry.Martin Karplus, Michael Levitt and Arieh Warshel received the 2013 Nobel Prize in Chemistry for "the development of multiscale models for complex chemical systems".

Fields of Application

The term *theoretical chemistry* may be defined as a mathematical description of chemistry, whereas *computational chemistry* is usually used when a mathematical method is sufficiently well developed that it can be automated for implementation on a computer. In theoretical chemistry, chemists, physicists, and mathematicians develop algorithms and computer programs to predict atomic and molecular properties and reaction paths for chemical reactions. Computational chemists, in contrast, may simply apply existing computer programs and methodologies to specific chemical questions.

Computational chemistry has two different aspects:

- Computational studies, used to find a starting point for a laboratory synthesis, or to assist in understanding experimental data, such as the position and source of spectroscopic peaks.

- Computational studies, used to predict the possibility of so far entirely unknown molecules or to explore reaction mechanisms not readily studied via experiments.

Thus, computational chemistry can assist the experimental chemist or it can challenge the experimental chemist to find entirely new chemical objects.

Several major areas may be distinguished within computational chemistry:

- The prediction of the molecular structure of molecules by the use of the simulation of forces, or more accurate quantum chemical methods, to find stationary points on the energy surface as the position of the nuclei is varied.

- Storing and searching for data on chemical entities.

- Identifying correlations between chemical structures and properties (*quantitative structure–property relationship* (QSPR) and *quantitative structure–activity relationship* (QSAR)).

- Computational approaches to help in the efficient synthesis of compounds.

- Computational approaches to design molecules that interact in specific ways with other molecules (e.g. drug design and catalysis).

Accuracy

The words *exact* and *perfect* do not appear here, as very few aspects of chemistry can be computed exactly. However, almost every aspect of chemistry can be described in a qualitative or approximate quantitative computational scheme.

Molecules consist of nuclei and electrons, so the methods of quantum mechanics apply. Computational chemists often attempt to solve the non-relativistic Schrödinger equation, with relativistic corrections added, although some progress has been made in solving the fully relativistic Dirac equation. In principle, it is possible to solve the Schrödinger equation in either its time-dependent or time-independent form, as appropriate for the problem in hand; in practice, this is not possible except for very small systems. Therefore, a great number of approximate methods strive to achieve the best trade-off between accuracy and computational cost.

Accuracy can always be improved with greater computational cost. Significant errors can present themselves in ab initio models comprising many electrons, due to the computational cost of full relativistic-inclusive methods. This complicates the study of molecules interacting with high atomic mass unit atoms, such as transitional metals and their catalytic properties. Present algorithms in computational chemistry can routinely calculate the properties of molecules that contain up to about 40 electrons with sufficient accuracy. Errors for energies can be less than a few kJ/mol. For geometries, bond lengths can be predicted within a few picometres and bond angles within 0.5 degrees. The treatment of larger molecules that contain a few dozen electrons is computationally tractable by approximate methods such as density functional theory (DFT).

There is some dispute within the field whether or not the latter methods are sufficient to describe complex chemical reactions, such as those in biochemistry. Large molecules can be studied by semi-empirical approximate methods. Even larger molecules are treated by classical mechanics methods that use what are called molecular mechanics (MM). In QM-MM methods, small parts of large complexes are treated quantum mechanically (QM), and the remainder is treated approximately (MM).

Methods

One molecular formula can represent more than one molecular isomer: a set of isomers. Each isomer is a local minimum on the energy surface (called the potential energy surface) created from the total energy (i.e., the electronic energy, plus the repulsion energy between the nuclei) as a function of the coordinates of all the nuclei. A stationary point is a geometry such that the derivative of the energy with respect to all displacements of the nuclei is zero. A local (energy) minimum is a stationary point where all such displacements lead to an increase in energy. The local minimum that is lowest is called the global minimum and corresponds to the most stable isomer. If there is one particular coordinate change that leads to a decrease in the total energy in both directions, the stationary point is a transition structure and the coordinate is the reaction coordinate. This process of determining stationary points is called geometry optimization.

The determination of molecular structure by geometry optimization became routine only after efficient methods for calculating the first derivatives of the energy with respect to all atomic coordinates became available. Evaluation of the related second derivatives allows the prediction of vibrational frequencies if harmonic motion is estimated. More importantly, it allows for the characterization of stationary points. The frequencies are related to the eigenvalues of the Hessian matrix, which contains second derivatives. If the eigenvalues are all positive, then the frequencies are all real and the stationary point is a local minimum. If one eigenvalue is negative (i.e., an imaginary frequency), then the stationary point is a transition structure. If more than one eigenvalue is negative, then the stationary point is a more complex one, and is usually of little interest. When one of these is found, it is necessary to move the search away from it if the experimenter is looking solely for local minima and transition structures.

The total energy is determined by approximate solutions of the time-dependent Schrödinger equation, usually with no relativistic terms included, and by making use of the Born–Oppenheimer approximation, which allows for the separation of electronic and nuclear motions, thereby simplifying the Schrödinger equation. This leads to the evaluation of the total energy as a sum of the electronic energy at fixed nuclei positions and the repulsion energy of the nuclei. A notable exception are certain approaches called direct quantum chemistry, which treat electrons and nuclei on a common footing. Density functional methods and semi-empirical methods are variants on the major theme. For very large systems, the relative total energies can be compared using molecular mechanics. The ways of determining the total energy to predict molecular structures are:

Ab Initio Methods

The programs used in computational chemistry are based on many different quantum-chemical methods that solve the molecular Schrödinger equation associated with the molecular Hamiltonian. Methods that do not include any empirical or semi-empirical parameters in their equations – being derived directly from theoretical principles, with no inclusion of experimental data – are called *ab initio methods*. This does not imply that the solution is an exact one; they are all approximate quantum mechanical calculations. It means that a particular approximation is rigorously defined on first principles (quantum theory) and then solved within an error margin that is qualitatively known beforehand. If numerical iterative methods must be used, the aim is to iterate until full machine accuracy is obtained (the best that is possible with a finite word length on the computer, and within the mathematical and/or physical approximations made).

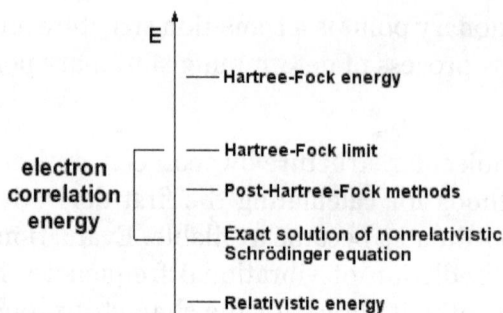

Diagram illustrating various *ab initio* electronic structure
methods in terms of energy. Spacings are not to scale.

The simplest type of *ab initio* electronic structure calculation is the Hartree–Fock method (HF), an extension of molecular orbital theory, in which the correlated electron-electron repulsion is not specifically taken into account; only its average effect is included in the calculation. As the basis set size is increased, the energy and wave function tend towards a limit called the Hartree–Fock limit. Many types of calculations (termed post-Hartree–Fock methods) begin with a Hartree–Fock calculation and subsequently correct for electron-electron repulsion, referred to also as electronic correlation. As these methods are pushed to the limit, they approach the exact solution of the non-relativistic Schrödinger equation. To obtain exact agreement with experiment, it is necessary to include relativistic and spin orbit terms, both of which are far more important for heavy atoms. In all of these approaches, along with choice of method, it is necessary to choose a basis set. This is a set of functions, usually centered on the different atoms in the molecule, which are used to expand the molecular orbitals with the linear combination of atomic orbitals (LCAO) molecular orbital method ansatz. Ab initio methods need to define a level of theory (the method) and a basis set.

The Hartree–Fock wave function is a single configuration or determinant. In some cases, particularly for bond breaking processes, this is inadequate, and several configurations must be used. Here, the coefficients of the configurations, and of the basis functions, are optimized together.

The total molecular energy can be evaluated as a function of the molecular geometry; in other words, the potential energy surface. Such a surface can be used for reaction dynamics. The stationary points of the surface lead to predictions of different isomers and the transition structures for conversion between isomers, but these can be determined without a full knowledge of the complete surface.

A particularly important objective, called computational thermochemistry, is to calculate thermochemical quantities such as the enthalpy of formation to chemical accuracy. Chemical accuracy is the accuracy required to make realistic chemical predictions and is generally considered to be 1 kcal/mol or 4 kJ/mol. To reach that accuracy in an economic way it is necessary to use a series of post-Hartree–Fock methods and combine the results. These methods are called quantum chemistry composite methods.

Density Functional Methods

Density functional theory (DFT) methods are often considered to be *ab initio methods* for determining the molecular electronic structure, even though many of the most common functionals use parameters derived from empirical data, or from more complex calculations. In DFT, the total energy is expressed in terms of the total one-electron density rather than the wave function. In this type of calculation, there is an approximate Hamiltonian and an approximate expression for the total electron density. DFT methods can be very accurate for little computational cost. Some methods combine the density functional exchange functional with the Hartree–Fock exchange term and are termed hybrid functional methods.

Semi-empirical and Empirical Methods

Semi-empirical quantum chemistry methods are based on the Hartree–Fock method formalism, but make many approximations and obtain some parameters from empirical data. They are very important in computational chemistry for treating large molecules where the full Hartree–Fock method without the approximations is too costly. The use of empirical parameters appears to allow some inclusion of correlation effects into the methods.

Semi-empirical methods follow what are often called empirical methods, where the two-electron part of the Hamiltonian is not explicitly included. For π-electron systems, this was the Hückel method proposed by Erich Hückel, and for all valence electron systems, the extended Hückel method proposed by Roald Hoffmann.

Molecular Mechanics

In many cases, large molecular systems can be modeled successfully while avoiding quantum mechanical calculations entirely. Molecular mechanics simulations, for example, use one classical expression for the energy of a compound, for instance the harmonic oscillator. All constants appearing in the equations must be obtained beforehand from experimental data or *ab initio* calculations.

The database of compounds used for parameterization, i.e., the resulting set of parameters and functions is called the force field, is crucial to the success of molecular mechanics calculations. A force field parameterized against a specific class of molecules, for instance proteins, would be expected to only have any relevance when describing other molecules of the same class.

These methods can be applied to proteins and other large biological molecules, and allow studies of the approach and interaction (docking) of potential drug molecules.

Methods for Solids

Computational chemical methods can be applied to solid state physics problems. The elec-

Definition and Scope

Principal and Mechanisms

Electrophysiology is the science and branch of physiology that pertains to the flow of ions (ion current) in biological tissues and, in particular, to the electrical recording techniques that enable the measurement of this flow. Classical electrophysiology techniques involve placing electrodes into various preparations of biological tissue. The principal types of electrodes are:

1. simple solid conductors, such as discs and needles (singles or arrays, often insulated except for the tip),

2. tracings on printed circuit boards, also insulated except for the tip, and

3. hollow tubes filled with an electrolyte, such as glass pipettes filled with potassium chloride solution or another electrolyte solution.

The principal preparations include:

1. living organisms,

2. excised tissue (acute or cultured),

3. dissociated cells from excised tissue (acute or cultured),

4. artificially grown cells or tissues, or

5. hybrids of the above.

If an electrode is small enough (micrometers) in diameter, then the electrophysiologist may choose to insert the tip into a single cell. Such a configuration allows direct observation and recording of the intracellular electrical activity of a single cell. However, at the same time such invasive setup reduces the life of the cell and causes a leak of substances across the cell membrane. Intracellular activity may also be observed using a specially formed (hollow) glass pipette containing an electrolyte. In this technique, the microscopic pipette tip is pressed against the cell membrane, to which it tightly adheres by an interaction between glass and lipids of the cell membrane. The electrolyte within the pipette may be brought into fluid continuity with the cytoplasm by delivering a pulse of negative pressure to the pipette in order to rupture the small patch of membrane encircled by the pipette rim (whole-cell recording). Alternatively, ionic continuity may be established by "perforating" the patch by allowing exogenous pore-forming agent within the electrolyte to insert themselves into the membrane patch (perforated patch recording). Finally, the patch may be left intact (patch recording).

The electrophysiologist may choose not to insert the tip into a single cell. Instead, the electrode tip may be left in continuity with the extracellular space. If the tip is small enough, such a configuration may allow indirect observation and recording of action potentials from a single cell, and is termed single-unit recording. Depending on the preparation and precise placement, an extracellular configuration may pick up the activity of several nearby cells simultaneously, and this is termed multi-unit recording.

As electrode size increases, the resolving power decreases. Larger electrodes are sensitive only to the net activity of many cells, termed local field potentials. Still larger electrodes, such as uninsulated needles and surface electrodes used by clinical and surgical neurophysiologists, are sensitive only to certain types of synchronous activity within populations of cells numbering in the millions.

Other classical electrophysiological techniques include single channel recording and amperometry.

Electrographic Modalities by Body Part

Electrophysiological recording in general is sometimes called electrography (from *electro-* + *-graphy*, "electrical recording"), with the record thus produced being an electrogram. However, the word *electrography* has other senses (including electrophotography), and the specific types of electrophysiological recording are usually called by specific names, constructed on the pattern of *electro-* + [body part combining form] + *-graphy* (abbreviation ExG). Relatedly, the word *electrogram* (not being needed for those other senses) often carries the specific meaning of intracardiac electrogram, which is like an electrocardiogram but with some invasive leads (inside the heart) rather than only noninvasive leads (on the skin). Electrophysiological recording for clinical diagnostic purposes is included within the category of electrodiagnostic testing. The various "ExG" modes are as follows:

Modality	Abbreviation	Body part	Common in clinical use
electrocardiography	ECG or EKG	heart (specifically, the cardiac muscle), with cutaneous electrodes (noninvasive)	Very common
electroatriography	EAG	atrial cardiac muscle	Uncommon
electroventriculography	EVG	ventricular cardiac muscle	Uncommon
intracardiac electrogram	EGM	heart (specifically, the cardiac muscle), with intracardiac electrodes (invasive)	Somewhat common
electroencephalography	EEG	brain (usually the cerebral cortex), with extracranial electrodes	Somewhat common
electrocorticography	ECoG or iEEG	brain (specifically the cerebral cortex), with intracranial electrodes	Somewhat common
electromyography	EMG	muscles throughout the body (usually skeletal, occasionally smooth)	Very common

electrooculography	EOG	eye (entire globe)	Somewhat common
electroretinography	ERG	retina specifically	Somewhat common
electronystagmography	ENG	eye via the corneoretinal potential	Somewhat common
electroolfactography	EOG	olfactory epithelium in mammals	Uncommon
electroantennography	EAG	olfactory receptors in arthropod antennae	Not applicable
electrocochleography	ECOG or ECochG	cochlea	Somewhat common
electrogastrography	EGG	stomach smooth muscle	Somewhat common
electrogastroenterography	EGEG	stomach and bowel smooth muscle	Somewhat common
electroglottography	EGG	glottis	Uncommon
electropalatography	EPG	palatal contact of tongue	Uncommon
electroarteriography	EAG	arterial flow via streaming potential detected through skin	Uncommon
electroblepharography	EBG	eyelid muscle	Uncommon
electrodermography	EDG	skin	Uncommon
electrohysterography	EHG	uterus	Uncommon
electroneuronography	ENeG or ENoG	nerves	Uncommon
electropneumography	EPG	lungs (chest movements)	Uncommon
electrospinography	ESG	spinal cord	Uncommon
electrovomerography	EVG	vomeronasal organ	Uncommon

Optical Electrophysiological Techniques

Optical electrophysiological techniques were created by scientists and engineers to overcome one of the main limitations of classical techniques. Classical techniques allow observation of electrical activity at approximately a single point within a volume of tissue. Essentially, classical techniques singularize a distributed phenomenon. Interest in the spatial distribution of bioelectric activity prompted development of molecules capable of emitting light in response to their electrical or chemical environment. Examples are voltage sensitive dyes and fluorescing proteins.

After introducing one or more such compounds into tissue via perfusion, injection or gene expression, the 1 or 2-dimensional distribution of electrical activity may be observed and recorded.

Intracellular Recording

Intracellular recording involves measuring voltage and/or current across the mem-

This technique was developed by Erwin Neher and Bert Sakmann who received the Nobel Prize in 1991. Conventional intracellular recording involves impaling a cell with a fine electrode; patch-clamp recording takes a different approach. A patch-clamp microelectrode is a micropipette with a relatively large tip diameter. The microelectrode is placed next to a cell, and gentle suction is applied through the microelectrode to draw a piece of the cell membrane (the 'patch') into the microelectrode tip; the glass tip forms a high resistance 'seal' with the cell membrane. This configuration is the "cell-attached" mode, and it can be used for studying the activity of the ion channels that are present in the patch of membrane. If more suction is now applied, the small patch of membrane in the electrode tip can be displaced, leaving the electrode sealed to the rest of the cell. This "whole-cell" mode allows very stable intracellular recording. A disadvantage (compared to conventional intracellular recording with sharp electrodes) is that the intracellular fluid of the cell mixes with the solution inside the recording electrode, and so some important components of the intracellular fluid can be diluted. A variant of this technique, the "perforated patch" technique, tries to minimise these problems. Instead of applying suction to displace the membrane patch from the electrode tip, it is also possible to make small holes on the patch with pore-forming agents so that large molecules such as proteins can stay inside the cell and ions can pass through the holes freely. Also the patch of membrane can be pulled away from the rest of the cell. This approach enables the membrane properties of the patch to be analysed pharmacologically.

Sharp Electrode Technique

In situations where one wants to record the potential inside the cell membrane with minimal effect on the ionic constitution of the intracellular fluid a sharp electrode can be used. These micropipettes (electrodes) are again like those for patch clamp pulled from glass capillaries, but the pore is much smaller so that there is very little ion exchange between the intracellular fluid and the electrolyte in the pipette. The resistance of the micropipette electrode is tens or hundreds of MΩ. Often the tip of the electrode is filled with various kinds of dyes like Lucifer yellow to fill the cells recorded from, for later confirmation of their morphology under a microscope. The dyes are injected by applying a positive or negative, DC or pulsed voltage to the electrodes depending on the polarity of the dye.

Extracellular Recording

Single-unit Recording

An electrode introduced into the brain of a living animal will detect electrical activity that is generated by the neurons adjacent to the electrode tip. If the electrode is a microelectrode, with a tip size of about 1 micrometre, the electrode will usually detect the activity of at most one neuron. Recording in this way is in general called "single-unit" recording. The action potentials recorded are very much like the action potentials that are recorded intracellularly, but the signals are very much smaller (typically about 1 mV). Most recordings of the activity of single neurons in anesthetized and conscious animals are made in this way. Recordings of single neurons in living animals have provided important insights into how the brain processes information. For example, David Hubel and Torsten Wiesel recorded the activity of single neurons in the primary visual cortex of the anesthetized cat, and showed how single neurons in this area respond to very specific features of a visual stimulus. Hubel and Wiesel were awarded the Nobel Prize in Physiology or Medicine in 1981.

Multi-unit Recording

If the electrode tip is slightly larger, then the electrode might record the activity generated by several neurons. This type of recording is often called "multi-unit recording", and is often used in conscious animals to record changes in the activity in a discrete brain area during normal activity. Recordings from one or more such electrodes that are closely spaced can be used to identify the number of cells around it as well as which of the spikes come from which cell. This process is called spike sorting and is suitable in areas where there are identified types of cells with well defined spike characteristics. If the electrode tip is bigger still, in general the activity of individual neurons cannot be distinguished but the electrode will still be able to record a field potential generated by the activity of many cells.

Field Potentials

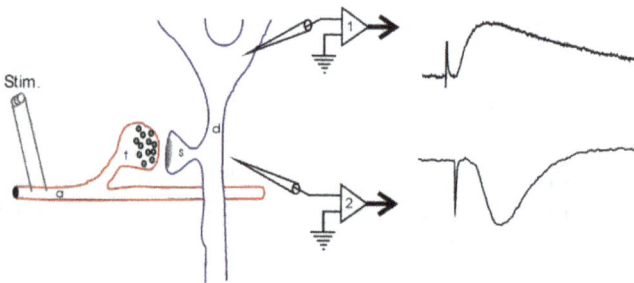

A schematic diagram showing a field potential recording from rat hippocampus. At the left is a schematic diagram of a presynaptic terminal and postsynaptic neuron. This is meant to represent a large population of synapses and neurons. When the synapse releases glutamate onto the postsynaptic cell, it opens ionotropic glutamate receptor channels. The net flow of current is inward, so a current sink is generated. A nearby electrode (#2) detects this as a negativity. An *intracellular* electrode placed inside the cell body (#1) records the change in membrane potential that the incoming current causes.

Extracellular field potentials are local current sinks or sources that are generated by the collective activity of many cells. Usually, a field potential is generated by the simultaneous activation of many neurons by synaptic transmission. The diagram to the right shows hippocampal synaptic field potentials. At the right, the lower trace shows a negative wave that corresponds to a current sink caused by positive charges entering cells through postsynaptic glutamate receptors, while the upper trace shows a positive wave that is generated by the current that leaves the cell (at the cell body) to complete the circuit. For more information, see local field potential.

Amperometry

Amperometry uses a carbon electrode to record changes in the chemical composition of the oxidized components of a biological solution. Oxidation and reduction is accomplished by changing the voltage at the active surface of the recording electrode in a process known as "scanning". Because certain brain chemicals lose or gain electrons at characteristic voltages, individual species can be identified. Amperometry has been used for studying exocytosis in the nervous and endocrine systems. Many monoamine neurotransmitters; e.g., norepinephrine (noradrenalin), dopamine, and serotonin (5-HT) are oxidizable. The method can also be used with cells that do not secrete oxidizable neurotransmitters by "loading" them with 5-HT or dopamine.

Planar Patch Clamp

Planar patch clamp is a novel method developed for high throughput electrophysiology. Instead of positioning a pipette on an adherent cell, cell suspension is pipetted on a chip containing a microstructured aperture.

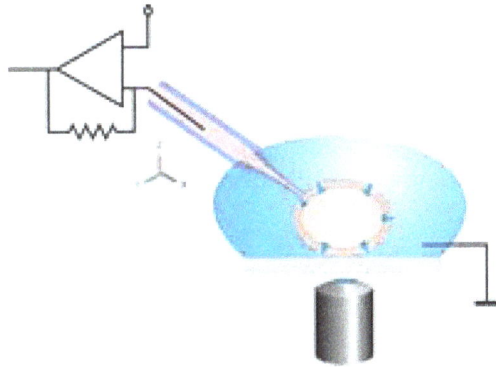

Schematic drawing of the classical patch clamp configuration. The patch pipette is moved to the cell using a micromanipulator under optical control. Relative movements between the pipette and the cell have to be avoided in order to keep the cell-pipette connection intact.

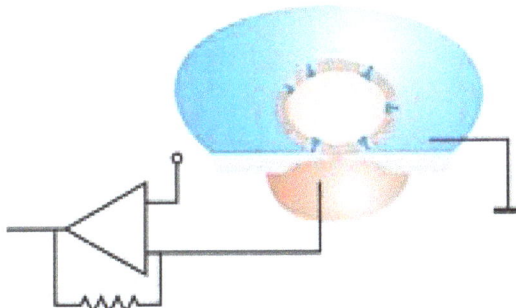

In planar patch configuration the cell is positioned by suction – relative movements between cell and aperture can then be excluded after sealing. An Antivibration table is not necessary.

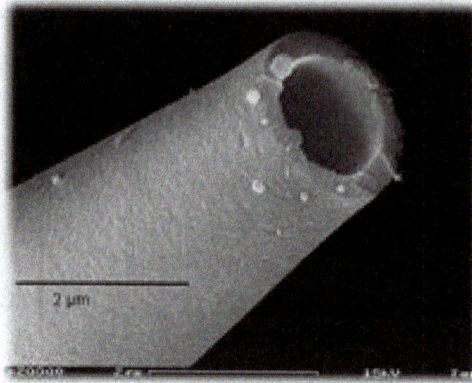

Scanning electron microscope image of a patch pipette

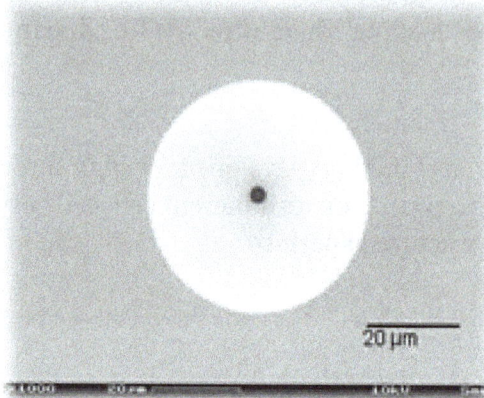

Scanning electron microscope image of a planar patch clamp chip.
Both the pipette and the chip are made from borosilicate glass.

A single cell is then positioned on the hole by suction and a tight connection (Gigaseal) is formed. The planar geometry offers a variety of advantages compared to the classical experiment:

- It allows for integration of microfluidics, which enables automatic compound application for ion channel screening.

- The system is accessible for optical or scanning probe techniques.

- Perfusion of the intracellular side can be performed.

Other Methods

Solid-supported Membrane (SSM)-Based

With this electrophysiological approach, proteoliposomes, membrane vesicles, or membrane fragments containing the channel or transporter of interest are adsorbed to a lipid monolayer painted over a functionalized electrode. This electrode consists of a glass support, a chromium layer, a gold layer, and an octadecyl mercaptane monolayer.

Because the painted membrane is supported by the electrode, it is called a solid-supported membrane. It is important to note that mechanical perturbations, which usually destroy a biological lipid membrane, do not influence the life-time of an SSM. The capacitive electrode (composed of the SSM and the absorbed vesicles) is so mechanically stable that solutions may be rapidly exchanged at its surface. This property allows the application of rapid substrate/ligand concentration jumps to investigate the electrogenic activity of the protein of interest, measured via capacitive coupling between the vesicles and the electrode.

Bioelectric Recognition Assay (BERA)

The bioelectric recognition assay (BERA) is a novel method for determination of various chemical and biological molecules by measuring changes in the membrane potential of cells immobilized in a gel matrix. Apart from the increased stability of the electrode-cell interface, immobilization preserves the viability and physiological functions of the cells. BERA is used primarily in biosensor applications in order to assay analytes that can interact with the immobilized cells by changing the cell membrane potential. In this way, when a positive sample is added to the sensor, a characteristic, "signature-like" change in electrical potential occurs. BERA is the core technology behind the recently launched pan-European FOODSCAN project, about pesticide and food risk assessment in Europe. BERA has been used for the detection of human viruses (hepatitis B and C viruses and herpes viruses), veterinary disease agents (foot and mouth disease virus, prions, and blue tongue virus), and plant viruses (tobacco and cucumber viruses) in a specific, rapid (1–2 minutes), reproducible, and cost-efficient fashion. The method has also been used for the detection of environmental toxins, such as pesticides and mycotoxins in food, and 2,4,6-trichloroanisole in cork and wine, as well as the determination of very low concentrations of the superoxide anion in clinical samples.

A BERA sensor has two parts:

- The consumable biorecognition elements

- The electronic read-out device with embedded artificial intelligence.

A recent advance is the development of a technique called molecular identification through membrane engineering (MIME). This technique allows for building cells with defined specificity for virtually any molecule of interest, by embedding thousands of artificial receptors into the cell membrane.

Computational Electrophysiology

While not strictly constituting an experimental measurement, methods have been developed to examine the conductive properties of proteins and biomembranes *in silico*. These are mainly molecular dynamics simulations in which a model system like a lipid bilayer is subjected to an externally applied voltage. Studies using these setups have been able to study dynamical phenomena like electroporation of membranes and ion translocation by channels.

The benefit of such methods is the high level of detail of the active conduction mechanism, given by the inherently high resolution and data density that atomistic simulation affords. There are significant drawbacks, given by the uncertainty of the legitimacy of the model and the computational cost of modeling systems that are large enough and over sufficient timescales to be considered reproducing the macroscopic properties of the systems themselves. While atomistic simulations may access timescales close to, or into the microsecond domain, this is still several orders of magnitude lower than even the resolution of experimental methods such as patch-clamping.

Clinical Reporting Guidelines

Minimum Information (MI) standards or reporting guidelines specify the minimum amount of meta data (information) and data required to meet a specific aim or aims in a clinical study. The "Minimum Information about a Neuroscience investigation" (MINI) family of reporting guideline documents aims to provide a consistent set of guidelines in order to report an electrophysiology experiment. In practice a MINI module comprises a checklist of information that should be provided (for example about the protocols employed) when a data set is described for publication.

Fluorescence spectroscopy

Fluorescence spectroscopy (also known as fluorometry or spectrofluorometry) is a type of electromagnetic spectroscopy that analyzes fluorescence from a sample. It involves using a beam of light, usually ultraviolet light, that excites the electrons in molecules of certain compounds and causes them to emit light; typically, but not necessarily, visible light. A complementary technique is absorption spectroscopy. In the special case of single molecule fluorescence spectroscopy, intensity fluctuations from the emitted light are measured from either single fluorophores, or pairs of fluorophores.

Devices that measure fluorescence are called fluorometers

Theory

Molecules have various states referred to as energy levels. Fluorescence spectroscopy is primarily concerned with electronic and vibrational states. Generally, the species being examined has a ground electronic state (a low energy state) of interest, and an excited electronic state of higher energy. Within each of these electronic states there are various vibrational states.

In fluorescence, the species is first excited, by absorbing a photon, from its ground electronic state to one of the various vibrational states in the excited electronic state. Collisions with other molecules cause the excited molecule to lose vibrational energy until it reaches the lowest vibrational state of the excited electronic state. This process is often visualized with a Jablonski diagram.

The molecule then drops down to one of the various vibrational levels of the ground electronic state again, emitting a photon in the process. As molecules may drop down into any of several vibrational levels in the ground state, the emitted photons will have different energies, and thus frequencies. Therefore, by analysing the different frequencies of light emitted in fluorescent spectroscopy, along with their relative intensities, the structure of the different vibrational levels can be determined.

For atomic species, the process is similar; however, since atomic species do not have vibrational energy levels, the emitted photons are often at the same wavelength as the incident radiation. This process of re-emitting the absorbed photon is "resonance fluorescence" and while it is characteristic of atomic fluorescence, is seen in molecular fluorescence as well.

In a typical fluorescence (emission) measurement, the excitation wavelength is fixed and the detection wavelength varies, while in a fluorescence excitation measurement the detection wavelength is fixed and the excitation wavelength is varied across a region of interest. An emission map is measured by recording the emission spectra resulting from a range of excitation wavelengths and combining them all together. This is a three dimensional surface data set: emission intensity as a function of excitation and emission wavelengths, and is typically depicted as a contour map.

Instrumentation

Two general types of instruments exist: filter fluorometers that use filters to isolate the incident light and fluorescent light and spectrofluorometers that use a diffraction gratingmonochromators to isolate the incident light and fluorescent light.

Both types use the following scheme: the light from an excitation source passes through a filter or monochromator, and strikes the sample. A proportion of the incident light is absorbed by the sample, and some of the molecules in the sample fluoresce. The fluorescent light is emitted in all directions. Some of this fluorescent light passes through a second filter or monochromator and reaches a detector, which is usually placed at 90° to the incident light beam to minimize the risk of transmitted or reflected incident light reaching the detector.

Various light sources may be used as excitation sources, including lasers, LED, and lamps; xenon arcs and mercury-vapor lamps in particular. A laser only emits light of high irradiance at a very narrow wavelength interval, typically under 0.01 nm, which makes an excitation monochromator or filter unnecessary. The disadvantage of this method is that the wavelength of a laser cannot be changed by much. A mercury vapor lamp is a line lamp, meaning it emits light near peak wavelengths. By contrast, a xenon arc has a continuous emission spectrum with nearly constant intensity in the range from 300-800 nm and a sufficient irradiance for measurements down to just above 200 nm.

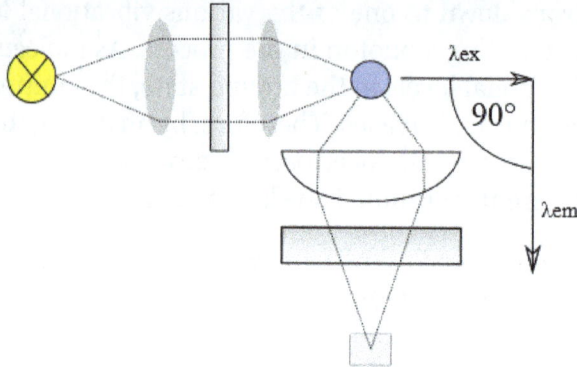

A simplistic design of the components of a fluorimeter

Filters and/or monochromators may be used in fluorimeters. A monochromator transmits light of an adjustable wavelength with an adjustable tolerance. The most common type of monochromator utilizes a diffraction grating, that is, collimated light illuminates a grating and exits with a different angle depending on the wavelength. The monochromator can then be adjusted to select which wavelengths to transmit. For allowing anisotropy measurements the addition of two polarization filters are necessary: One after the excitation monochromator or filter, and one before the emission monochromator or filter.

As mentioned before, the fluorescence is most often measured at a 90° angle relative to the excitation light. This geometry is used instead of placing the sensor at the line of the excitation light at a 180° angle in order to avoid interference of the transmitted excitation light. No monochromator is perfect and it will transmit some stray light, that is, light with other wavelengths than the targeted. An ideal monochromator would only transmit light in the specified range and have a high wavelength-independent transmission. When measuring at a 90° angle, only the light scattered by the sample causes stray light. This results in a better signal-to-noise ratio, and lowers the detection limit by approximately a factor 10000, when compared to the 180° geometry. Furthermore, the fluorescence can also be measured from the front, which is often done for turbid or opaque samples .

The detector can either be single-channeled or multichanneled. The single-channeled detector can only detect the intensity of one wavelength at a time, while the multichanneled detects the intensity of all wavelengths simultaneously, making the emission monochromator or filter unnecessary. The different types of detectors have both advantages and disadvantages.

The most versatile fluorimeters with dual monochromators and a continuous excitation light source can record both an excitation spectrum and a fluorescence spectrum. When measuring fluorescence spectra, the wavelength of the excitation light is kept constant, preferably at a wavelength of high absorption, and the emission monochromator scans the spectrum. For measuring excitation spectra, the wavelength passing though the emission filter or monochromator is kept constant and the excitation monochromator is scanning. The excitation spectrum generally is identical to the absorption spectrum as the fluorescence intensity is proportional to the absorption.

Analysis of Data

At low concentrations the fluorescenceintensity will generally be proportional to the concentration of the fluorophore.

Unlike in UV/visible spectroscopy, 'standard', device independent spectra are not easily attained. Several factors influence and distort the spectra, and corrections are necessary to attain 'true', i.e. machine-independent, spectra. The different types of distortions will here be classified as being either instrument- or sample-related. Firstly, the distortion arising from the instrument is discussed. As a start, the light source intensity and wavelength characteristics varies over time during each experiment and between each experiment. Furthermore, no lamp has a constant intensity at all wavelengths. To correct this, a beam splitter can be applied after the excitation monochromator or filter to direct a portion of the light to a reference detector.

Additionally, the transmission efficiency of monochromators and filters must be taken into account. These may also change over time. The transmission efficiency of the monochromator also varies depending on wavelength. This is the reason that an optional reference detector should be placed after the excitation monochromator or filter. The percentage of the fluorescence picked up by the detector is also dependent upon the system. Furthermore, the detector quantum efficiency, that is, the percentage of photons detected, varies between different detectors, with wavelength and with time, as the detector inevitably deteriorates.

Two other topics that must be considered include the optics used to direct the radiation and the means of holding or containing the sample material (called a cuvette or cell). For most UV, visible, and NIR measurements the use of precision quartz cuvettes is necessary. In both cases, it is important to select materials that have relatively little absorption in the wavelength range of interest. Quartz is ideal because it transmits from 200 nm-2500 nm; higher grade quartz can even transmit up to 3500 nm, whereas the absorption properties of other materials can mask the fluorescence from the sample.

Correction of all these instrumental factors for getting a 'standard' spectrum is a tedious process, which is only applied in practice when it is strictly necessary. This is the case when measuring the quantum yield or when finding the wavelength with the highest emission intensity for instance.

As mentioned earlier, distortions arise from the sample as well. Therefore, some aspects of the sample must be taken into account too. Firstly, photodecomposition may decrease the intensity of fluorescence over time. Scattering of light must also be taken into account. The most significant types of scattering in this context are Rayleigh and Raman scattering. Light scattered by Rayleigh scattering has the same wavelength as the incident light, whereas in Raman scattering the scattered light changes wavelength usually to longer wavelengths. Raman scattering is the result of a virtual electronic state induced by the excitation light. From this virtual state, the molecules may relax back to a vibrational level other than the vibrational ground state. In fluorescence spectra, it is always seen at a constant wavenumber difference relative to the excitation wavenumber e.g. the peak appears at a wavenumber 3600 cm^{-1} lower than the excitation light in water.

Other aspects to consider are the inner filter effects. These include reabsorption. Reabsorption happens because another molecule or part of a macromolecule absorbs at the wavelengths at which the fluorophore emits radiation. If this is the case, some or all of the photons emitted by the fluorophore may be absorbed again. Another inner filter effect occurs because of high concentrations of absorbing molecules, including the fluorophore. The result is that the intensity of the excitation light is not constant throughout the solution. Resultingly, only a small percentage of the excitation light reaches the fluorophores that are visible for the detection system. The inner filter effects change the spectrum and intensity of the emitted light and they must therefore be considered when analysing the emission spectrum of fluorescent light.

Tryptophan Fluorescence

The fluorescence of a folded protein is a mixture of the fluorescence from individual aromatic residues. Most of the intrinsic fluorescence emissions of a folded protein are due to excitation of tryptophan residues, with some emissions due to tyrosine and phenylalanine; but disulfide bonds also have appreciable absorption in this wavelength range. Typically, tryptophan has a wavelength of maximum absorption of 280 nm and an emission peak that is solvatochromic, ranging from ca. 300 to 350 nm depending in the polarity of the local environment Hence, protein fluorescence may be used as a diagnostic of the conformational state of a protein. Furthermore, tryptophan fluorescence is strongly influenced by the proximity of other residues (*i.e.*, nearby *protonated* groups such as Asp or Glu can cause quenching of Trp fluorescence). Also, energy transfer between tryptophan and the other fluorescent amino acids is possible, which would affect the analysis, especially in cases where the Förster acidic approach is taken. In addition, tryptophan is a relatively rare amino acid; many proteins contain only one or a few tryptophan residues. Therefore, tryptophan fluorescence can be a very sensitive measurement of the conformational state of individual tryptophan residues. The advantage compared to extrinsic probes is that the protein itself is not changed. The use of intrinsic fluorescence for the study of protein conformation is in practice limited to cases with few (or perhaps only one) tryptophan residues, since each experiences a different local environment, which gives rise to different emission spectra.

Tryptophan is an important intrinsic fluorescent probe (amino acid), which can be used to estimate the nature of microenvironment of the tryptophan. When performing experiments with denaturants, surfactants or other amphiphilic molecules, the microenvironment of the tryptophan might change. For example, if a protein containing a single tryptophan in its 'hydrophobic' core is denatured with increasing temperature, a red-shifted emission spectrum will appear. This is due to the exposure of the tryptophan to an aqueous environment as opposed to a hydrophobic protein interior. In contrast, the addition of a surfactant to a protein which contains a tryptophan which is exposed to the aqueous solvent will cause a blue-shifted emission spectrum if the tryptophan is embedded in the surfactant vesicle or micelle. Proteins that lack tryptophan may be coupled to a fluorophore.

With fluorescence excitation at 295 nm, the tryptophan emission spectrum is dominant over the weaker tyrosine and phenylalanine fluorescence.

Applications

Fluorescence spectroscopy is used in, among others, biochemical, medical, and chemical research fields for analyzing organic compounds. There has also been a report of its use in differentiating malignant skin tumors from benign.

Atomic Fluorescence Spectroscopy (AFS) techniques are useful in other kinds of analysis/measurement of a compound present in air or water, or other media, such as CVAFS which is used for heavy metals detection, such as mercury.

Fluorescence can also be used to redirect photons.

Additionally, Fluorescence spectroscopy can be adapted to the microscopic level using microfluorimetry

In analytical chemistry, fluorescence detectors are used with HPLC.

Microscopy

Microscopic examination in an biochemical laboratory

Microscopy is the technical field of using microscopes to view objects and areas of objects that cannot be seen with the naked eye (objects that are not within the resolution range of the normal eye). There are three well-known branches of microscopy: optical, electron, and scanning probe microscopy.

Scanning electron microscope image of pollen

Optical and electron microscopy involve the diffraction, reflection, or refraction of electromagnetic radiation/electron beams interacting with the specimen, and the collection of the scattered radiation or another signal in order to create an image. This process may be carried out by wide-field irradiation of the sample (for example standard light microscopy and transmission electron microscopy) or by scanning of a fine beam over the sample (for example confocal laser scanning microscopy and scanning electron microscopy). Scanning probe microscopy involves the interaction of a scanning probe with the surface of the object of interest. The development of microscopy revolutionized biology, gave rise to the field of histology and so remains an essential technique in the life and physical sciences.

Optical Microscopy

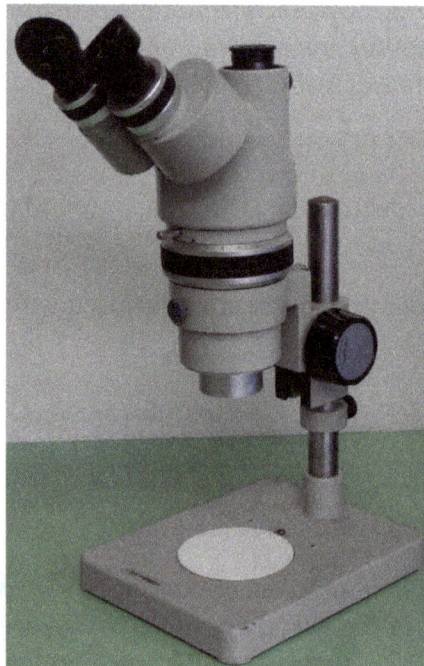

Stereo microscope

Optical or light microscopy involves passing visible light transmitted through or reflected from the sample through a single or multiple lenses to allow a magnified view of the sample. The resulting image can be detected directly by the eye, imaged on a photographic plate or captured digitally. The single lens with its attachments, or the system of lenses and imaging equipment, along with the appropriate lighting equipment, sample stage and support, makes up the basic light microscope. The most recent development is the digital microscope, which uses a CCD camera to focus on the exhibit of interest. The image is shown on a computer screen, so eye-pieces are unnecessary.

Limitations

Limitations of standard optical microscopy (bright field microscopy) lie in three areas;

- This technique can only image dark or strongly refracting objects effectively.

- Diffraction limits resolution to approximately 0.2 micrometres. This limits the practical magnification limit to ~1500x.

- Out of focus light from points outside the focal plane reduces image clarity.

Live cells in particular generally lack sufficient contrast to be studied successfully, since the internal structures of the cell are colourless and transparent. The most common way to increase contrast is to stain the different structures with selective dyes, but this often involves killing and fixing the sample. Staining may also introduce artifacts, apparent structural details that are caused by the processing of the specimen and are thus not legitimate features of the specimen. In general, these techniques make use of differences in the refractive index of cell structures. It is comparable to looking through a glass window: you (bright field microscopy) don't see the glass but merely the dirt on the glass. There is a difference, as glass is a denser material, and this creates a difference in phase of the light passing through. The human eye is not sensitive to this difference in phase, but clever optical solutions have been thought out to change this difference in phase into a difference in amplitude (light intensity).

Techniques

In order to improve specimen contrast or highlight certain structures in a sample special techniques must be used. A huge selection of microscopy techniques are available to increase contrast or label a sample.

- Four examples of transillumination techniques used to generate contrast in a sample of tissue paper. 1.559 µm/pixel.

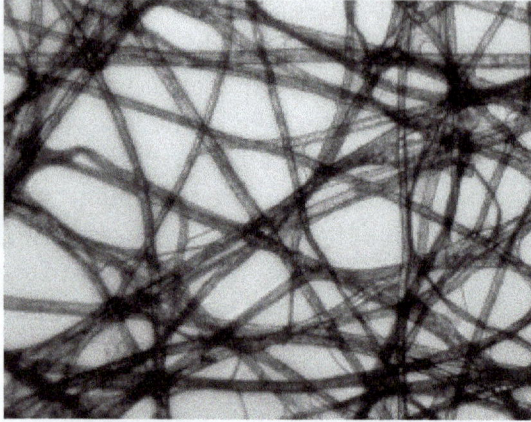

Bright field illumination, sample contrast comes from absorbance of light in the sample.

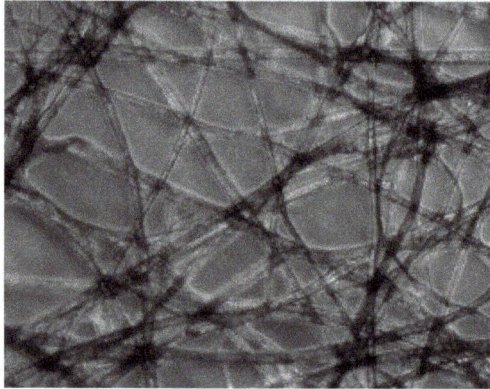

Phase contrast illumination, sample contrast comes from interference
of different path lengths of light through the sample.

Bright Field

Bright field microscopy is the simplest of all the light microscopy techniques. Sample illumination is via transmitted white light, i.e. illuminated from below and observed from above. Limitations include low contrast of most biological samples and low apparent resolution due to the blur of out of focus material. The simplicity of the technique and the minimal sample preparation required are significant advantages.

Oblique Illumination

The use of oblique (from the side) illumination gives the image a 3-dimensional appearance and can highlight otherwise invisible features. A more recent technique based on this method is *Hoffmann's modulation contrast*, a system found on inverted microscopes for use in cell culture. Oblique illumination suffers from the same limitations as bright field microscopy (low contrast of many biological samples; low apparent resolution due to out of focus objects).

Dark Field

Dark field microscopy is a technique for improving the contrast of unstained, transparent specimens. Dark field illumination uses a carefully aligned light source to minimize the quantity of directly transmitted (unscattered) light entering the image plane, collecting only the light scattered by the sample. Dark field can dramatically improve image contrast – especially of transparent objects – while requiring little equipment setup or sample preparation. However, the technique suffers from low light intensity in final image of many biological samples, and continues to be affected by low apparent resolution.

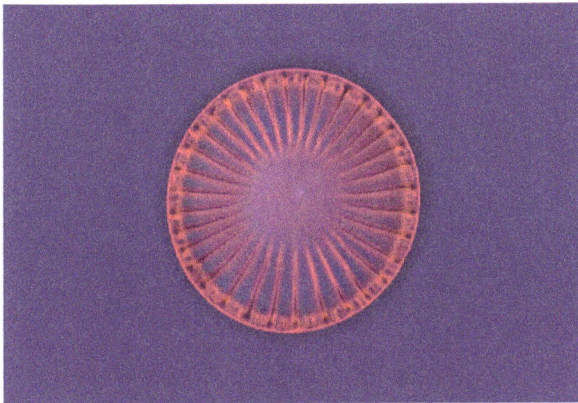

A diatom under Rheinberg illumination

Rheinberg illumination is a special variant of dark field illumination in which transparent, colored filters are inserted just before the condenser so that light rays at high aperture are differently colored than those at low aperture (i.e. the background to the specimen may be blue while the object appears self-luminous red). Other color combinations are possible but their effectiveness is quite variable.

Dispersion Staining

Dispersion staining is an optical technique that results in a colored image of a colorless object. This is an optical staining technique and requires no stains or dyes to produce a color effect. There are five different microscope configurations used in the broader technique of dispersion staining. They include brightfield Becke line, oblique, darkfield, phase contrast, and objective stop dispersion staining.

Phase Contrast

More sophisticated techniques will show proportional differences in optical density. Phase contrast is a widely used technique that shows differences in refractive index as difference in contrast. It was developed by the Dutch physicist Frits Zernike in the 1930s (for which he was awarded the Nobel Prize in 1953). The nucleus in a cell for example will show up darkly against the surrounding cytoplasm. Contrast is excellent;

however it is not for use with thick objects. Frequently, a halo is formed even around small objects, which obscures detail. The system consists of a circular annulus in the condenser, which produces a cone of light. This cone is superimposed on a similar sized ring within the phase-objective. Every objective has a different size ring, so for every objective another condenser setting has to be chosen. The ring in the objective has special optical properties: it, first of all, reduces the direct light in intensity, but more importantly, it creates an artificial phase difference of about a quarter wavelength. As the physical properties of this direct light have changed, interference with the diffracted light occurs, resulting in the phase contrast image. One disadvantage of phase-contrast microscopy is halo formation (halo-light ring).

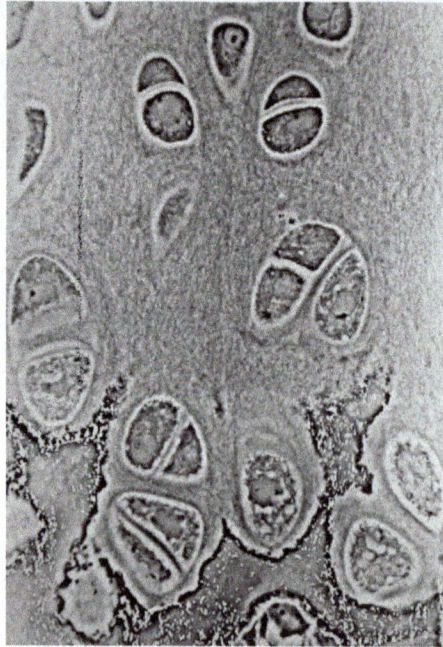

Phase-contrast light micrograph of hyaline cartilage showing chondrocytes and organelles, lacunae and extracellular matrix.

Differential Interference Contrast

Superior and much more expensive is the use of interference contrast. Differences in optical density will show up as differences in relief. A nucleus within a cell will actually show up as a globule in the most often used differential interference contrast system according to Georges Nomarski. However, it has to be kept in mind that this is an *optical effect*, and the relief does not necessarily resemble the true shape. Contrast is very good and the condenser aperture can be used fully open, thereby reducing the depth of field and maximizing resolution.

The system consists of a special prism (Nomarski prism, Wollaston prism) in the condenser that splits light in an ordinary and an extraordinary beam. The spatial difference between the two beams is minimal (less than the maximum resolution of the objective).

After passage through the specimen, the beams are reunited by a similar prism in the objective.

In a homogeneous specimen, there is no difference between the two beams, and no contrast is being generated. However, near a refractive boundary (say a nucleus within the cytoplasm), the difference between the ordinary and the extraordinary beam will generate a relief in the image. Differential interference contrast requires a polarized light source to function; two polarizing filters have to be fitted in the light path, one below the condenser (the polarizer), and the other above the objective (the analyzer).

Note: In cases where the optical design of a microscope produces an appreciable lateral separation of the two beams we have the case of classical interference microscopy, which does not result in relief images, but can nevertheless be used for the quantitative determination of mass-thicknesses of microscopic objects.

Interference Reflection

An additional technique using interference is interference reflection microscopy (also known as reflected interference contrast, or RIC). It relies on cell adhesion to the slide to produce an interference signal. If there is no cell attached to the glass, there will be no interference.

Interference reflection microscopy can be obtained by using the same elements used by DIC, but without the prisms. Also, the light that is being detected is reflected and not transmitted as it is when DIC is employed.

Fluorescence

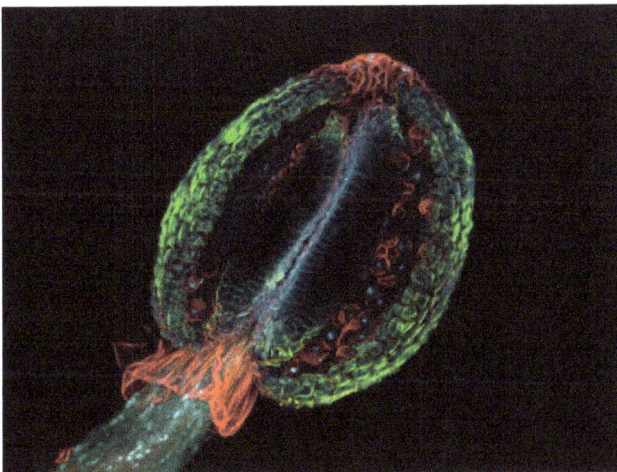

Images may also contain artifacts. This is a confocal laser scanningfluorescencemicrograph of thale cress anther (part of stamen). The picture shows among other things a nice red flowing collar-like structure just below the anther. However, an intact thale cress stamen does not have such collar, this is a fixation artifact: the stamen has been cut below the picture frame, and epidermis (upper layer of cells) of stamen stalk has peeled off, forming a non-characteristic structure. Photo: Heiti Paves from Tallinn University of Technology.

When certain compounds are illuminated with high energy light, they emit light of a lower frequency. This effect is known as fluorescence. Often specimens show their characteristic autofluorescence image, based on their chemical makeup.

This method is of critical importance in the modern life sciences, as it can be extremely sensitive, allowing the detection of single molecules. Many different fluorescent dyes can be used to stain different structures or chemical compounds. One particularly powerful method is the combination of antibodies coupled to a fluorophore as in immunostaining. Examples of commonly used fluorophores are fluorescein or rhodamine.

The antibodies can be tailor-made for a chemical compound. For example, one strategy often in use is the artificial production of proteins, based on the genetic code (DNA). These proteins can then be used to immunize rabbits, forming antibodies which bind to the protein. The antibodies are then coupled chemically to a fluorophore and used to trace the proteins in the cells under study.

Highly efficient fluorescent proteins such as the green fluorescent protein (GFP) have been developed using the molecular biology technique of gene fusion, a process that links the expression of the fluorescent compound to that of the target protein. This combined fluorescent protein is, in general, non-toxic to the organism and rarely interferes with the function of the protein under study. Genetically modified cells or organisms directly express the fluorescently tagged proteins, which enables the study of the function of the original protein in vivo.

Growth of protein crystals results in both protein and salt crystals. Both are colorless and microscopic. Recovery of the protein crystals requires imaging which can be done by the intrinsic fluorescence of the protein or by using transmission microscopy. Both methods require an ultraviolet microscope as protein absorbs light at 280 nm. Protein will also fluorescence at approximately 353 nm when excited with 280 nm light.

Since fluorescence emission differs in wavelength (color) from the excitation light, an ideal fluorescent image shows only the structure of interest that was labeled with the fluorescent dye. This high specificity led to the widespread use of fluorescence light microscopy in biomedical research. Different fluorescent dyes can be used to stain different biological structures, which can then be detected simultaneously, while still being specific due to the individual color of the dye.

To block the excitation light from reaching the observer or the detector, filter sets of high quality are needed. These typically consist of an excitation filter selecting the range of excitation wavelengths, a dichroic mirror, and an emission filter blocking the excitation light. Most fluorescence microscopes are operated in the Epi-illumination mode (illumination and detection from one side of the sample) to further decrease the amount of excitation light entering the detector.

An example of fluorescence microscopy today is two-photon or multi-photon imaging. Two photon imaging allows imaging of living tissues up to a very high depth by enabling greater excitation light penetration and reduced background emission signal.

Confocal

Confocal microscopy uses a scanning point of light and a pinhole to prevent out of focus light from reaching the detector. Compared to full sample illumination, confocal microscopy gives slightly higher resolution, and significantly improves optical sectioning. Confocal microscopy is, therefore, commonly used where 3D structure is important.

Single Plane Illumination Microscopy and Light Sheet Fluorescence Microscopy

Using a plane of light formed by focusing light through a cylindrical lens at a narrow angle or by scanning a line of light in a plane perpendicular to the axis of objective, high resolution optical sections can be taken. Single plane illumination, or light sheet illumination, is also accomplished using beam shaping techniques incorporating multiple-prism beam expanders. The images are captured by CCDs. These variants allow very fast and high signal to noise ratio image capture.

Wide-field Multiphoton Microscopy

Wide-field multiphoton microscopy refers to an optical non-linear imaging technique tailored for ultrafast imaging in which a large area of the object is illuminated and imaged without the need for scanning. High intensities are required to induce non-linear optical processes such as two-photon fluorescence or second harmonic generation. In scanning multiphoton microscopes the high intensities are achieved by tightly focusing the light, and the image is obtained by stage- or beam-scanning the sample. In wide-field multiphoton microscopy the high intensities are best achieved using an optically amplified pulsed laser source to attain a large field of view (~100 μm). The image in this case is obtained as a single frame with a CCD without the need of scanning, making the technique particularly useful to visualize dynamic processes simultaneously across the object of interest. With wide-field multiphoton microscopy the frame rate can be increased up to a 1000-fold compared to multiphoton scanning microscopy.

Deconvolution

Fluorescence microscopy is a powerful technique to show specifically labeled structures within a complex environment and to provide three-dimensional information of biological structures. However, this information is blurred by the fact that, upon illumination, all fluorescently labeled structures emit light, irrespective of whether they are in focus or not. So an image of a certain structure is always blurred by the contribution of light from structures that are out of focus. This phenomenon results in a loss of contrast especially when using objectives with a high resolving power, typically oil immersion objectives with a high numerical aperture.

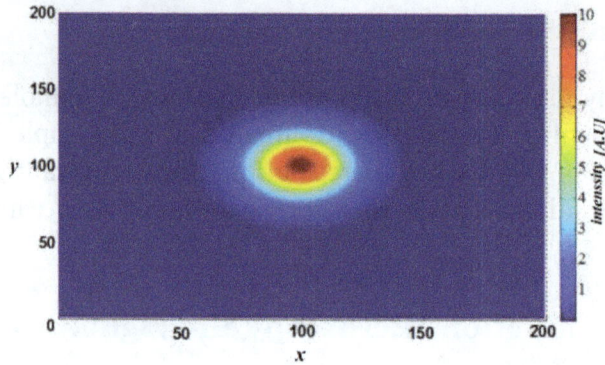

Mathematically modeled Point Spread Function of a pulsed THz laser imaging system.

However, blurring is not caused by random processes, such as light scattering, but can be well defined by the optical properties of the image formation in the microscope imaging system. If one considers a small fluorescent light source (essentially a bright spot), light coming from this spot spreads out further from our perspective as the spot becomes more out of focus. Under ideal conditions, this produces an "hourglass" shape of this point source in the third (axial) dimension. This shape is called the point spread function (PSF) of the microscope imaging system. Since any fluorescence image is made up of a large number of such small fluorescent light sources, the image is said to be "convolved by the point spread function". The mathematically modeled PSF of a terahertz laser pulsed imaging system is shown on the right.

The output of an imaging system can be described using the equation:

$$s(x, y) = PSF(x, y) * o(x, y) + n$$

Where n is the additive noise. Knowing this point spread function means that it is possible to reverse this process to a certain extent by computer-based methods commonly known as deconvolution microscopy. There are various algorithms available for 2D or 3D deconvolution. They can be roughly classified in *nonrestorative* and *restorative* methods. While the nonrestorative methods can improve contrast by removing out-of-focus light from focal planes, only the restorative methods can actually reassign light to its proper place of origin. Processing fluorescent images in this manner can be an advantage over directly acquiring images without out-of-focus light, such as images from confocal microscopy, because light signals otherwise eliminated become useful information. For 3D deconvolution, one typically provides a series of images taken from different focal planes (called a Z-stack) plus the knowledge of the PSF, which can be derived either experimentally or theoretically from knowing all contributing parameters of the microscope.

Sub-diffraction Techniques

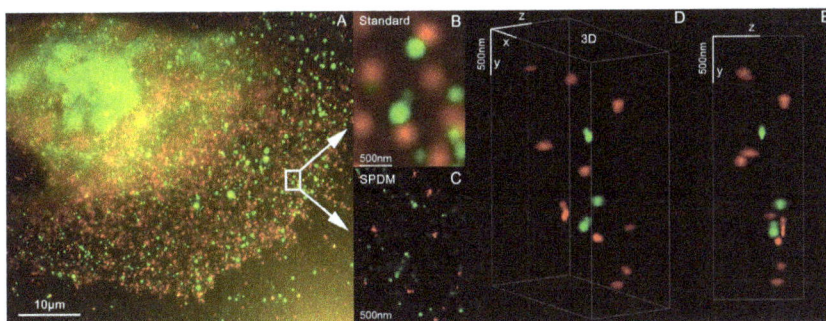

Example of super-resolution microscopy. Image of Her3 and Her2, target of the breast cancer drugTrastuzumab, within a cancer cell.

A multitude of super-resolution microscopy techniques have been developed in recent times which circumvent the diffraction barrier.

This is mostly achieved by imaging a sufficiently static sample multiple times and either modifying the excitation light or observing stochastic changes in the image.

Knowledge of and chemical control over fluorophore photophysics is at the core of these techniques, by which resolutions of ~20 nanometers are regularly obtained.

Serial Time-encoded Amplified Microscopy

Serial time encoded amplified microscopy (STEAM) is an imaging method that provides ultrafast shutter speed and frame rate, by using optical image amplification to circumvent the fundamental trade-off between sensitivity and speed, and a single-pixel photodetector to eliminate the need for a detector array and readout time limitations The method is at least 1000 times faster than the state-of-the-art CCD and CMOS cameras. Consequently, it is potentially useful for a broad range of scientific, industrial, and biomedical applications that require high image acquisition rates, including real-time diagnosis and evaluation of shockwaves, microfluidics, MEMS, and laser surgery.

Extensions

Most modern instruments provide simple solutions for micro-photography and image recording electronically. However such capabilities are not always present and the more experienced microscopist will, in many cases, still prefer a hand drawn image to a photograph. This is because a microscopist with knowledge of the subject can accurately convert a three-dimensional image into a precise two-dimensional drawing. In a photograph or other image capture system however, only one thin plane is ever in good focus.

The creation of careful and accurate micrographs requires a microscopical technique using a monocular eyepiece. It is essential that both eyes are open and that the eye that is not observing down the microscope is instead concentrated on a sheet of paper on the bench besides the microscope. With practice, and without moving the head or eyes, it is possible to accurately record the observed details by tracing round the observed shapes by simultaneously "seeing" the pencil point in the microscopical image.

Practicing this technique also establishes good general microscopical technique. It is always less tiring to observe with the microscope focused so that the image is seen at infinity and with both eyes open at all times.

Other Enhancements

Microspectroscopy:spectroscopy with a microscope

X-ray

As resolution depends on the wavelength of the light. Electron microscopy has been developed since the 1930s that use electron beams instead of light. Because of the much smaller wavelength of the electron beam, resolution is far higher.

Though less common, X-ray microscopy has also been developed since the late 1940s. The resolution of X-ray microscopy lies between that of light microscopy and electron microscopy.

Electron Microscopy

Until the invention of sub-diffraction microscopy, the wavelength of the light limited the resolution of traditional microscopy to around 0.2 micrometers. In order to gain higher resolution, the use of an electron beam with a far smaller wavelength is used in electron microscopes.

- Transmission electron microscopy (TEM) is quite similar to the compound light microscope, by sending an electron beam through a very thin slice of the specimen. The resolution limit in 2005 was around 0.05 nanometer and has not increased appreciably since that time.

- Scanning electron microscopy (SEM) visualizes details on the surfaces of specimens and gives a very nice 3D view. It gives results much like those of the stereo light microscope. The best resolution for SEM in 2011 was 0.4 nanometer.

Electron microscopes equipped for X-ray spectroscopy can provide qualitative and quantitative elemental analysis.

Scanning Probe Microscopy

This is a sub-diffraction technique. Examples of scanning probe microscopes are the atomic force microscope (AFM), the Scanning tunneling microscope, the photonic force microscope and the recurrence tracking microscope. All such methods use the physical contact of a solid probe tip to scan the surface of an object, which is supposed to be almost flat.

Ultrasonic Force

Ultrasonic Force Microscopy (UFM) has been developed in order to improve the details and image contrast on "flat" areas of interest where AFM images are limited in contrast. The combination of AFM-UFM allows a near field acoustic microscopic image to be generated. The AFM tip is used to detect the ultrasonic waves and overcomes the limitation of wavelength that occurs in acoustic microscopy. By using the elastic changes under the AFM tip, an image of much greater detail than the AFM topography can be generated.

Ultrasonic force microscopy allows the local mapping of elasticity in atomic force microscopy by the application of ultrasonic vibration to the cantilever or sample. In an attempt to analyze the results of ultrasonic force microscopy in a quantitative fashion, a force-distance curve measurement is done with ultrasonic vibration applied to the cantilever base, and the results are compared with a model of the cantilever dynamics and tip-sample interaction based on the finite-difference technique.

Ultraviolet Microscopy

Ultraviolet microscopes have two main purposes. The first is to utilize the shorter wavelength of ultraviolet electromagnetic energy to improve the image resolution beyond that of the diffraction limit of standard optical microscopes. This technique is used for non-destructive inspection of devices with very small features such as those found in modern semiconductors. The second application for UV microscopes is contrast enhancement where the response of individual samples is enhanced, relative to their surrounding, due to the interaction of light with the molecules within the sample itself. One example is in the growth of protein crystals. Protein crystals are formed in salt solutions. As salt and protein crystals are both formed in the growth process, and both are commonly transparent to the human eye, they cannot be differentiated with a standard optical microscope. As the tryptophan of protein absorbs light at 280 nm, imaging with a UV microscope with 280 nm bandpass filters makes it simple to differentiate between the two types of crystals. The protein crystals appear dark while the salt crystals are transparent.

Infrared Microscopy

The term *infrared microscopy* refers to microscopy performed at infrared wavelengths. In the typical instrument configuration a Fourier Transform Infrared Spectrometer (FTIR) is combined with an optical microscope and an infrared detector. The infrared detector can be a single point detector, a linear array or a 2D focal plane array. The FTIR provides the ability to perform chemical analysis via infrared spectroscopy and the microscope and point or array detector enable this chemical analysis to be spatially resolved, i.e. performed at different regions of the sample. As such, the technique is also called infrared microspectroscopy (an alternative architecture involves the combination of a tuneable infrared light source and single point detector on a flying objective). This technique is frequently used for infrared chemical imaging, where the image contrast is determined by the response of individual sample regions to particular IR wavelengths selected by the user, usually specific IR absorption bands and associated molecular resonances . A key limitation of conventional infrared microspectroscopy is that the spatial resolution is diffraction-limited. Specifically the spatial resolution is limited to a figure related to the wavelength of the light. For practical IR microscopes, the spatial resolution is limited to 1-3X the wavelength, depending on the specific technique and instrument used. For mid-IR wavelengths, this sets a practical spatial resolution limit of ~3-30 µm.

IR versions of sub-diffraction microscopy also exist . These include IR NSOM,photothermal microspectroscopy, and atomic force microscope based infrared spectroscopy (AFM-IR).

Digital Holographic Microscopy

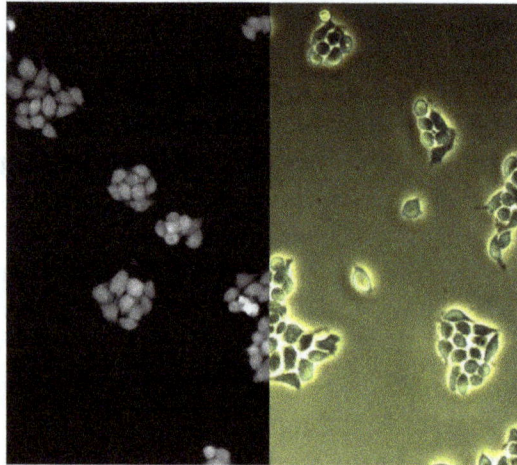

Human cells imaged by DHM phase shift (left) and phase contrast microscopy (right).

In digital holographic microscopy (DHM), interfering wave fronts from a coherent (monochromatic) light-source are recorded on a sensor. The image is digitally reconstructed by a computer from the recorded hologram. Besides the ordinary bright field image, a phase shift image is created.

DHM can operate both in reflection and transmission mode. In reflection mode, the phase shift image provides a relative distance measurement and thus represents a topography map of the reflecting surface. In transmission mode, the phase shift image provides a label-free quantitative measurement of the optical thickness of the specimen. Phase shift images of biological cells are very similar to images of stained cells and have successfully been analyzed by high content analysis software.

A unique feature of DHM is the ability to adjust focus after the image is recorded, since all focus planes are recorded simultaneously by the hologram. This feature makes it possible to image moving particles in a volume or to rapidly scan a surface. Another attractive feature is DHM's ability to use low cost optics by correcting optical aberrations by software.

Digital Pathology (Virtual Microscopy)

Digital pathology is an image-based information environment enabled by computer technology that allows for the management of information generated from a digital slide. Digital pathology is enabled in part by virtual microscopy, which is the practice of converting glass slides into digital slides that can be viewed, managed, and analyzed.

Laser Microscopy

Laser microscopy is a rapidly growing field that uses laser illumination sources in various forms of microscopy. For instance, laser microscopy focused on biological applications uses ultrashort pulse lasers, in a number of techniques labeled as nonlinear microscopy, saturation microscopy, and two-photon excitation microscopy.

High-intensity, short-pulse laboratory x-ray lasers have been under development for several years. When this technology comes to fruition, it will be possible to obtain magnified three-dimensional images of elementary biological structures in the living state at a precisely defined instant. For optimum contrast between water and protein and for best sensitivity and resolution, the laser should be tuned near the nitrogen line at about 0.3 nanometers. Resolution will be limited mainly by the hydrodynamic expansion that occurs while the necessary number of photons is being registered. Thus, while the specimen is destroyed by the exposure, its configuration can be captured before it explodes.

Scientists have been working on practical designs and prototypes for x-ray holographic microscopes, despite the prolonged development of the appropriate laser.

Amateur Microscopy

Amateur Microscopy is the investigation and observation of biological and non-biological specimens for recreational purposes. Collectors of minerals, insects, seashells, and plants may use microscopes as tools to uncover features that help them classify their collected items. Other amateurs may be interested in observing the life found in pond water and of other samples. Microscopes may also prove useful for the water quality assessment for people that keep a home aquarium. Photographic documentation and drawing of the microscopic images are additional tasks that augment the spectrum of tasks of the amateur. There are even competitions for photomicrograph art. Participants of this pastime may either use commercially prepared microscopic slides or engage in the task of specimen preparation.

While microscopy is a central tool in the documentation of biological specimens, it is, in general, insufficient to justify the description of a new species based on microscopic investigations alone. Often genetic and biochemical tests are necessary to confirm the discovery of a new species. A laboratory and access to academic literature is a necessity, which is specialized and, in general, not available to amateurs. There is, however, one huge advantage that amateurs have above professionals: time to explore their surroundings. Often, advanced amateurs team up with professionals to validate their findings and (possibly) describe new species.

In the late 1800s, amateur microscopy became a popular hobby in the United States and Europe. Several 'professional amateurs' were being paid for their sampling trips and microscopic explorations by philanthropists, to keep them amused on the Sunday afternoon (e.g., the diatom specialist A. Grunow, being paid by (among others) a Belgian industrialist). Professor John Phin published "Practical Hints on the Selection and Use of the Microscope (Second Edition, 1878)," and was also the editor of the "American Journal of Microscopy."

Examples of Amateur Microscopy Images:

"house bee" Mouth 100X

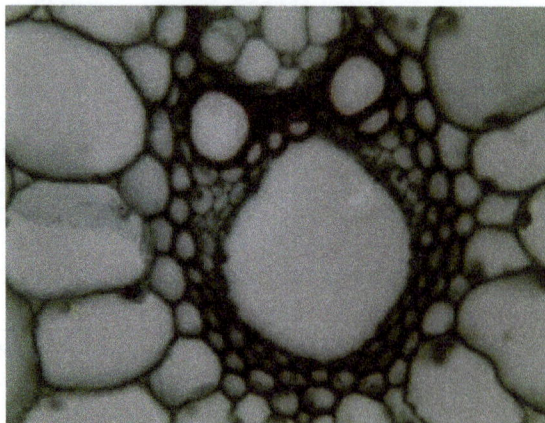

Rice Stem cs 400X

Nuclear Magnetic Resonance Spectroscopy

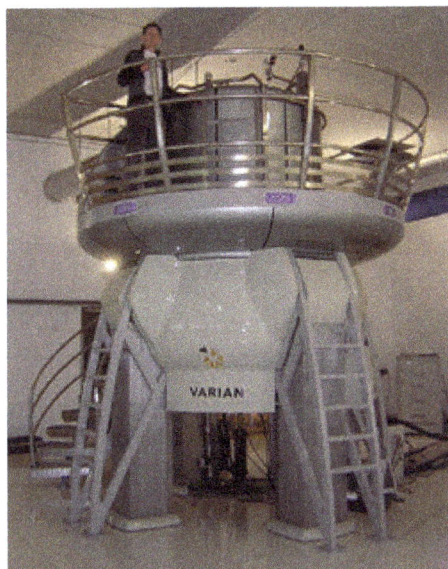

A 900MHz NMR instrument with a 21.1 T magnet at HWB-NMR, Birmingham, UK

Nuclear magnetic resonance spectroscopy, most commonly known as NMR spectroscopy, is a research technique that exploits the magnetic properties of certain atomic nuclei. This type of spectroscopy determines the physical and chemical properties of atoms or the molecules in which they are contained. It relies on the phenomenon of nuclear magnetic resonance and can provide detailed information about the structure, dynamics, reaction state, and chemical environment of molecules. The intramolecular magnetic field around an atom in a molecule changes the resonance frequency, thus giving access to details of the electronic structure of a molecule and its individual functional groups.

Most frequently, NMR spectroscopy is used by chemists and biochemists to investigate the properties of organic molecules, although it is applicable to any kind of sample that contains nuclei possessing spin. Suitable samples range from small compounds analyzed with 1-dimensional proton or carbon-13 NMR spectroscopy to large proteins or nucleic acids using 3 or 4-dimensional techniques. The impact of NMR spectroscopy on the sciences has been substantial because of the range of information and the diversity of samples, including solutions and solids.

NMR spectra are unique, well-resolved, analytically tractable and often highly predictable for small molecules. Thus, in organic chemistry practice, NMR analysis is used to confirm the identity of a substance. Different functional groups are obviously distinguishable, and identical functional groups with differing neighboring substituents still give distinguishable signals. NMR has largely replaced traditional wet chemistry tests such as color reagents or typical chromatography for identification. A disadvantage is that a relatively large amount, 2–50 mg, of a purified substance is required, although it may be recovered through a workup. Preferably, the sample should be dissolved in a solvent, because NMR analysis of solids requires a dedicated MAS machine and may not give equally well-resolved spectra. The timescale of NMR is relatively long, and thus it is not suitable for observing fast phenomena, producing only an averaged spectrum. Although large amounts of impurities do show on an NMR spectrum, better methods exist for detecting impurities, as NMR is inherently not very sensitive - though at higher frequencies, sensitivity narrows.

NMR spectrometers are relatively expensive; universities usually have them, but they are less common in private companies. Modern NMR spectrometers have a very strong, large and expensive liquid helium-cooled superconducting magnet, because resolution directly depends on magnetic field strength. Less expensive machines using permanent magnets and lower resolution are also available, which still give sufficient performance for certain application such as reaction monitoring and quick checking of samples. There are even benchtop NMR spectrometers.

History

The Purcell group at Harvard University and the Bloch group at Stanford University independently developed NMR spectroscopy in the late 1940s and early 1950s. Edward Mills Purcell and Felix Bloch shared the 1952 Nobel Prize in Physics for their discoveries.

Basic NMR Techniques

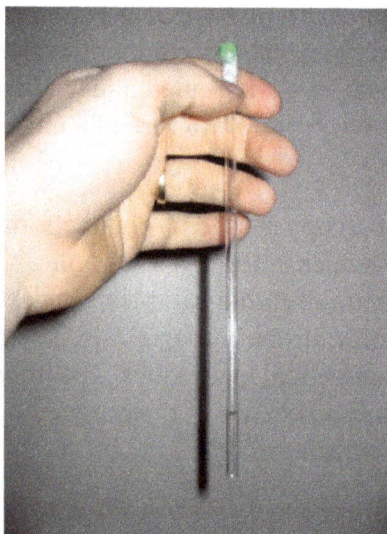

The NMR sample is prepared in a thin-walled glass tube - an NMR tube.

Resonant Frequency

When placed in a magnetic field, NMR active nuclei (such as ^{1}H or ^{13}C) absorb electromagnetic radiation at a frequency characteristic of the isotope. The resonant frequency, energy of the absorption, and the intensity of the signal are proportional to the strength of the magnetic field. For example, in a 21 Tesla magnetic field, protons resonate at 900 MHz. It is common to refer to a 21 T magnet as a 900 MHz magnet, although different nuclei resonate at a different frequency at this field strength in proportion to their nuclear magnetic moments.

Sample Handling

A NMR spectrometer typically consists of a spinning sample-holder inside a very strong magnet, a radio-frequency emitter and a receiver with a probe (an antenna assembly) that goes inside the magnet to surround the sample, optionally gradient coils for diffusion measurements and electronics to control the system. Spinning the sample is necessary to average out diffusional motion. Whereas, measurements of diffusion constants (*diffusion ordered spectroscopy* or DOSY) are done the sample stationary and spinning off, and flow cells can be used for online analysis of process flows.

Deuterated Solvents

The vast majority of nuclei in a solution would belong to the solvent, and most regular solvents are hydrocarbons and would contain NMR-reactive protons. Thus, deuterium (hydrogen-2) is substituted (99+%). The most used deuterated solvent is deuterochloroform ($CDCl_3$), although deuterium oxide (D_2O) and deuterated DMSO (DMSO-d_6) are used for hydrophilic analytes. The chemical shifts are slightly different in different solvents, depending on electronic solvation effects. NMR spectra are often calibrated against the known solvent residual proton peak instead of added tetramethylsilane.

Shim and Lock

To detect the very small frequency shifts due to nuclear magnetic resonance, the applied magnetic field must be constant throughout the sample volume. High resolution NMR spectrometers use shims to adjust the homogeneity of the magnetic field to parts per billion (ppb) in a volume of a few cubic centimeters. In order to detect and compensate for inhomogeneity and drift in the magnetic field, the spectrometer maintains a "lock" on the solvent deuterium frequency with a separate lock unit. The operator usually has to optimize the shim parameters manually to obtain the best possible intensity and prevent artifacts.

Acquisition of Spectra

Upon excitation of the sample with radio frequency (60–1000 MHz) pulse, a nuclear magnetic resonance response - a free induction decay (FID) - is obtained. It is a very weak signal, and requires sensitive radio receivers to pick up. A Fourier transform is done to extract the frequency-domain spectrum from the raw time-domain FID. A spectrum from a single FID has a low signal-to-noise ratio, but fortunately it improves readily with averaging of repeated acquisitions. Good ^1H NMR spectra can be acquired with 16 repeats, which takes only minutes. However, for heavier elements than hydrogen, the relaxation time is rather long, e.g. around 8 seconds for ^{13}C. Thus, acquisition of quantitative heavy-element spectra can be time-consuming, taking tens of minutes to hours.

If the second excitation pulse is sent prematurely before the relaxation is complete, the average magnetization vector still points in a nonparallel direction, giving suboptimal absorption and emission of the pulse. In practice, the peak areas are then not proportional to the stoichiometry; only the presence, but not the amount of functional groups is possible to discern. An invension recovery experiment can be done to determine the relaxation time and thus the required delay between pulses. A 180° pulse, an adjustable delay, and a 90° pulse is transmitted. When the 90° pulse exactly cancels out the signal, the delay corresponds to the time needed for 90° of relaxation. Inversion recovery is worthwhile for quantitive ^{13}C, 2D and other time-consuming experiments.

Chemical Shift

A spinning charge generates a magnetic field that results in a magnetic moment proportional to the spin. In the presence of an external magnetic field, two spin states exist (for a spin 1/2 nucleus): one spin up and one spin down, where one aligns with the magnetic field and the other opposes it. The difference in energy (ΔE) between the two spin states increases as the strength of the field increases, but this difference is usually very small, leading to the requirement for strong NMR magnets (1-20 T for modern NMR instruments). Irradiation of the sample with energy corresponding to the exact spin state separation of a specific set of nuclei will cause excitation of those set of nuclei in the lower energy state to the higher energy state.

For spin 1/2 nuclei, the energy difference between the two spin states at a given magnetic field strength is proportional to their magnetic moment. However, even if all protons have the same magnetic moments, they do not give resonant signals at the same frequency values. This difference arises from the differing electronic environments of the nucleus of interest. Upon application of an external magnetic field, these electrons move in response to the field and generate local magnetic fields that oppose the much stronger applied field. This local field thus "shields" the proton from the applied magnetic field, which must therefore be increased in order to achieve resonance (absorption of rf energy). Such increments are very small, usually in parts per million (ppm). For instance, the proton peak from an aldehyde is shifted ca. 10 ppm compared to a hydrocarbon peak, since as an electron-withdrawing group, the carbonyl deshields the proton by reducing the local electron density. The difference between 2.3487 T and 2.3488 T is therefore about 42 ppm. However a frequency scale is commonly used to designate the NMR signals, even though the spectrometer may operate by sweeping the magnetic field, and thus the 42 ppm is 4200 Hz for a 100 MHz reference frequency (rf).

However given that the location of different NMR signals is dependent on the external magnetic field strength and the reference frequency, the signals are usually reported relative to a reference signal, usually that of TMS (tetramethylsilane). Additionally, since the distribution of NMR signals is field dependent, these frequencies are divided by the spectrometer frequency. However, since we are dividing Hz by MHz, the resulting number would be too small, and thus it is multiplied by a million. This operation therefore gives a locator number called the "chemical shift" with units of parts per million. In general, chemical shifts for protons are highly predictable since the shifts are primarily determined by simpler shielding effects (electron density), but the chemical shifts for many heavier nuclei are more strongly influenced by other factors including excited states ("paramagnetic" contribution to shielding tensor).

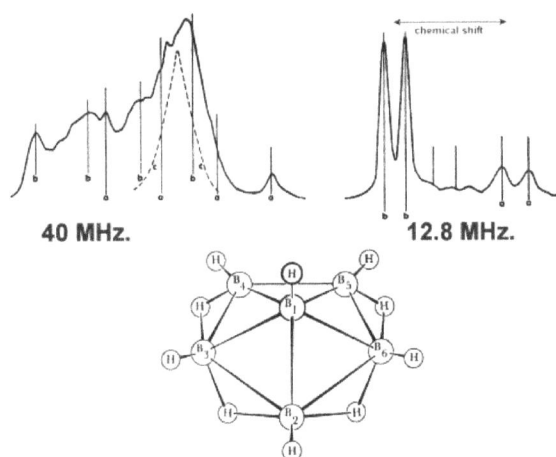

Example of the chemical shift: NMR spectrum of hexaborane B_6H_{10} showing peaks shifted in frequency, which give clues as to the molecular structure. (click to read interpretation details)

The chemical shift provides information about the structure of the molecule. The conversion of the raw data to this information is called *assigning* the spectrum. For example, for the ¹H-NMR spectrum for ethanol (CH_3CH_2OH), one would expect signals at each of three specific chemical shifts: one for the CH_3 group, one for the CH_2 group and one for the OH group. A typical CH_3 group has a shift around 1 ppm, a CH_2 attached to an OH has a shift of around 4 ppm and an OH has a shift anywhere from 2–6 ppm depending on the solvent used and the amount of hydrogen bonding. While the O atom does draw electron density away from the attached H through their mutual sigma bond, the electron lone pairs on the O bathe the H in their shielding effect.

In Paramagnetic NMR spectroscopy, measurements are conducted on paramagnetic samples. The paramagnetism gives rise to very diverse chemical shifts. In 1H NMR spectroscopy, the chemical shift range can span 500 ppm.

Because of molecular motion at room temperature, the three methyl protons *average out* during the NMR experiment (which typically requires a few ms). These protons become degenerate and form a peak at the same chemical shift.

The shape and area of peaks are indicators of chemical structure too. In the example above—the proton spectrum of ethanol—the CH_3 peak has three times the area as the OH peak. Similarly the CH_2 peak would be twice the area of the OH peak but only 2/3 the area of the CH_3 peak.

Software allows analysis of signal intensity of peaks, which under conditions of optimal relaxation, correlate with the number of protons of that type. Analysis of signal intensity is done by integration—the mathematical process that calculates the area under a curve. The analyst must integrate the peak and not measure its height because the peaks also have *width*—and thus its size is dependent on its area not its height. However, it should be mentioned that the number of protons, or any other observed nucleus, is only proportional to the intensity, or the integral, of the NMR signal in the very simplest one-dimensional NMR experiments. In more elaborate experiments, for instance, experiments typically used to obtain carbon-13 NMR spectra, the integral of the signals depends on the relaxation rate of the nucleus, and its scalar and dipolar coupling constants. Very often these factors are poorly known - therefore, the integral of the NMR signal is very difficult to interpret in more complicated NMR experiments.

J-coupling

Multiplicity	Intensity Ratio
Singlet (s)	1
Doublet (d)	1:1
Triplet (t)	1:2:1
Quartet (q)	1:3:3:1
Quintet	1:4:6:4:1
Sextet	1:5:10:10:5:1
Septet	1:6:15:20:15:6:1

Some of the most useful information for structure determination in a one-dimensional NMR spectrum comes from J-coupling or scalar coupling (a special case of spin-spin coupling) between NMR active nuclei. This coupling arises from the interaction of different spin states through the chemical bonds of a molecule and results in the splitting of NMR signals. These splitting patterns can be complex or simple and, likewise, can be straightforwardly interpretable or deceptive. This coupling provides detailed insight into the connectivity of atoms in a molecule.

Ethanol

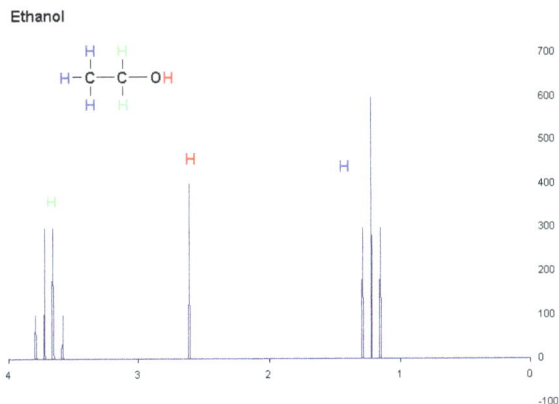

Example ^{1}H NMR spectrum (1-dimensional) of ethanol plotted as signal intensity vs. chemical shift. There are three different types of H atoms in ethanol regarding NMR. The hydrogen (H) on the -OH group is not coupling with the other H atoms and appears as a singlet, but the CH_3- and the -CH_2- hydrogens are coupling with each other, resulting in a triplet and quartet respectively.

Coupling to n equivalent (spin ½) nuclei splits the signal into a $n+1$ multiplet with intensity ratios following Pascal's triangle as described on the right. Coupling to additional spins will lead to further splittings of each component of the multiplet e.g. coupling to two different spin ½ nuclei with significantly different coupling constants will lead to a *doublet of doublets* (abbreviation: dd). Note that coupling between nuclei that are chemically equivalent (that is, have the same chemical shift) has no effect on the NMR spectra and couplings between nuclei that are distant (usually more than 3 bonds apart for protons in flexible molecules) are usually too small to cause observable splittings. *Long-range* couplings over more than three bonds can often be observed in cyclic and aromatic compounds, leading to more complex splitting patterns.

For example, in the proton spectrum for ethanol described above, the CH_3 group is split into a *triplet* with an intensity ratio of 1:2:1 by the two neighboring CH_2 protons. Similarly, the CH_2 is split into a *quartet* with an intensity ratio of 1:3:3:1 by the three neighboring CH_3 protons. In principle, the two CH_2 protons would also be split again into a *doublet* to form a *doublet of quartets* by the hydroxyl proton, but intermolecular exchange of the acidic hydroxyl proton often results in a loss of coupling information.

Coupling to any spin ½ nuclei such as phosphorus-31 or fluorine-19 works in this fashion (although the magnitudes of the coupling constants may be very different). But the splitting patterns differ from those described above for nuclei with spin greater than ½ because the spin quantum number has more than two possible values. For instance, coupling to deuterium (a spin 1 nucleus) splits the signal into a *1:1:1 triplet* because the spin 1 has three spin states. Similarly, a spin 3/2 nucleus splits a signal into a *1:1:1:1 quartet* and so on.

Coupling combined with the chemical shift (and the integration for protons) tells us not only about the chemical environment of the nuclei, but also the number of *neighboring* NMR active nuclei within the molecule. In more complex spectra with multiple peaks at similar chemical shifts or in spectra of nuclei other than hydrogen, coupling is often the only way to distinguish different nuclei.

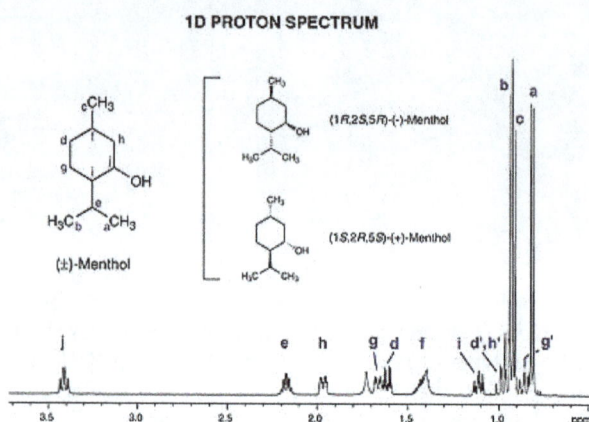

1D PROTON SPECTRUM

^{1}H NMR spectrum of menthol with chemical shift in ppm on the horizontal axis. Each magnetically inequivalent proton has a characteristic shift, and couplings to other protons appear as splitting of the peaks into multiplets: e.g. peak *a*, because of the three magnetically equivalent protons in methyl group *a*, couple to one adjacent proton (*e*) and thus appears as a doublet.

Second-order (or strong) Coupling

The above description assumes that the coupling constant is small in comparison with the difference in NMR frequencies between the inequivalent spins. If the shift separation decreases (or the coupling strength increases), the multiplet intensity patterns are first distorted, and then become more complex and less easily analyzed (especially if more than two spins are involved). Intensification of some peaks in a multiplet is achieved at the expense of the remainder, which sometimes almost disappear in the background noise, although the integrated area under the peaks remains constant. In most high-field NMR, however, the distortions are usually modest and the characteristic distortions (*roofing*) can in fact help to identify related peaks.

Some of these patterns can be analyzed with the method published by John Pople, though it has limited scope.

Second-order effects decrease as the frequency difference between multiplets increases, so that high-field (i.e. high-frequency) NMR spectra display less distortion than lower frequency spectra. Early spectra at 60 MHz were more prone to distortion than spectra from later machines typically operating at frequencies at 200 MHz or above.

Magnetic Inequivalence

More subtle effects can occur if chemically equivalent spins (i.e., nuclei related by symmetry and so having the same NMR frequency) have different coupling relationships to external spins. Spins that are chemically equivalent but are not indistinguishable (based on their coupling relationships) are termed magnetically inequivalent. For example, the 4 H sites of 1,2-dichlorobenzene divide into two chemically equivalent pairs by symmetry, but an individual member of one of the pairs has different couplings to the spins making up the other pair. Magnetic inequivalence can lead to highly complex spectra which can only be analyzed by computational modeling. Such effects are more common in NMR spectra of aromatic and other non-flexible systems, while conformational averaging about C-C bonds in flexible molecules tends to equalize the couplings between protons on adjacent carbons, reducing problems with magnetic inequivalence.

Correlation Spectroscopy

Correlation spectroscopy is one of several types of two-dimensional nuclear magnetic resonance (NMR) spectroscopy or 2D-NMR. This type of NMR experiment is best known by its acronym, COSY. Other types of two-dimensional NMR include J-spectroscopy, exchange spectroscopy (EXSY), Nuclear Overhauser effect spectroscopy (NOESY), total correlation spectroscopy (TOCSY) and heteronuclear correlation experiments, such as HSQC, HMQC, and HMBC. In correlation spectroscopy, emission is centered on an the peak of an individual nucleus; if its magnetic field is correlated with another nucleus by through-bond (COSY, HSQC, etc.) or through-space (NOE) coupling, a response can also be detected on the frequency of the correlated nucleus. Two-dimensional NMR spectra provide more information about a molecule than one-dimensional NMR spectra and are especially useful in determining the structure of a molecule, particularly for molecules that are too complicated to work with using one-dimensional NMR. The first two-dimensional experiment, COSY, was proposed by Jean Jeener, a professor at Université Libre de Bruxelles, in 1971. This experiment was later implemented by Walter P. Aue, Enrico Bartholdi and Richard R. Ernst, who published their work in 1976.

Solid-state Nuclear Magnetic Resonance

A variety of physical circumstances do not allow molecules to be studied in solution, and at the same time not by other spectroscopic techniques to an atomic level, either. In solid-phase media, such as crystals, microcrystalline powders, gels, anisotropic solutions, etc., it is in particular the dipolar coupling and chemical shift anisotropy that become dominant to the behaviour of the nuclear spin systems. In conventional solution-state NMR spectroscopy, these additional interactions would lead to a significant broadening of spectral lines. A variety of techniques allows establishing high-resolution conditions, that can, at least for ^{13}C spectra, be comparable to solution-state NMR spectra.

Two important concepts for high-resolution solid-state NMR spectroscopy are the limitation of possible molecular orientation by sample orientation, and the reduction of anisotropic nuclear magnetic interactions by sample spinning. Of the latter approach, fast spinning around the magic angle is a very prominent method, when the system comprises spin 1/2 nuclei. Spinning rates of ca. 20 kHz are used, which demands special equipment. A number of intermediate techniques, with samples of partial alignment or reduced mobility, is currently being used in NMR spectroscopy.

Applications in which solid-state NMR effects occur are often related to structure investigations on membrane proteins, protein fibrils or all kinds of polymers, and chemical analysis in inorganic chemistry, but also include "exotic" applications like the plant leaves and fuel cells. For example, Rahmani et al. studied the effect of pressure and temperature on the bicellar structures' self-assembly using deuterium NMR spectroscopy.

Biomolecular NMR Spectroscopy

Proteins

Much of the innovation within NMR spectroscopy has been within the field of protein-NMR spectroscopy, an important technique in structural biology. A common goal of these investigations is to obtain high resolution 3-dimensional structures of the protein, similar to what can be achieved by X-ray crystallography. In contrast to X-ray crystallography, NMR spectroscopy is usually limited to proteins smaller than 35 kDa, although larger structures have been solved. NMR spectroscopy is often the only way to obtain high resolution information on partially or wholly intrinsically unstructured proteins. It is now a common tool for the determination of Conformation Activity Relationships where the structure before and after interaction with, for example, a drug candidate is compared to its known biochemical activity. Proteins are orders of magnitude larger than the small organic molecules discussed earlier in this article, but the basic NMR techniques and some NMR theory also applies. Because of the much higher number of atoms present in a protein molecule in comparison with a small organic compound, the basic 1D spectra become crowded with overlapping signals to an extent where direct spectral analysis becomes untenable. Therefore, multidimensional (2, 3 or 4D) experiments have been devised to deal with this problem. To facilitate these experiments, it is desirable to isotopically label the protein with ^{13}C and ^{15}N because the predominant naturally occurring isotope ^{12}C is not NMR-active and the nuclear quadrupole moment of the predominant naturally occurring ^{14}N isotope prevents high resolution information from being obtained from this nitrogen isotope. The most important method used for structure determination of proteins utilizes NOE experiments to measure distances between pairs of atoms within the molecule. Subsequently, the distances obtained are used to generate a 3D structure of the molecule by solving a distance geometry problem. NMR can also be used to obtain information on the dynamics and conformational flexibility of different regions of a protein.

Nucleic Acids

"Nucleic acid NMR" is the use of NMR spectroscopy to obtain information about the structure and dynamics of polynucleic acids, such as DNA or RNA. As of 2003, nearly half of all known RNA structures had been determined by NMR spectroscopy.

Nucleic acid and protein NMR spectroscopy are similar but differences exist. Nucleic acids have a smaller percentage of hydrogen atoms, which are the atoms usually observed in NMR spectroscopy, and because nucleic acid double helices are stiff and roughly linear, they do not fold back on themselves to give "long-range" correlations. The types of NMR usually done with nucleic acids are ^1H or proton NMR, ^{13}C NMR, ^{15}N NMR, and ^{31}P NMR. Two-dimensional NMR methods are almost always used, such as correlation spectroscopy (COSY) and total coherence transfer spectroscopy (TOCSY) to detect through-bond nuclear couplings, and nuclear Overhauser effect spectroscopy (NOESY) to detect couplings between nuclei that are close to each other in space.

Parameters taken from the spectrum, mainly NOESY cross-peaks and coupling constants, can be used to determine local structural features such as glycosidic bond angles, dihedral angles (using the Karplus equation), and sugar pucker conformations. For large-scale structure, these local parameters must be supplemented with other structural assumptions or models, because errors add up as the double helix is traversed, and unlike with proteins, the double helix does not have a compact interior and does not fold back upon itself. NMR is also useful for investigating nonstandard geometries such as bent helices, non-Watson–Crick basepairing, and coaxial stacking. It has been especially useful in probing the structure of natural RNA oligonucleotides, which tend to adopt complex conformations such as stem-loops and pseudoknots. NMR is also useful for probing the binding of nucleic acid molecules to other molecules, such as proteins or drugs, by seeing which resonances are shifted upon binding of the other molecule.

Carbohydrates

Carbohydrate NMR spectroscopy addresses questions on the structure and conformation of carbohydrates.

Electron Microscope

A modern transmission electron microscope

Electron microscope constructed by Ernst Ruska in 1933

An electron microscope is a microscope that uses a beam of accelerated electrons as a source of illumination. As the wavelength of an electron can be up to 100,000 times shorter than that of visible light photons, electron microscopes have a higher resolving power than light microscopes and can reveal the structure of smaller objects. A transmission electron microscope can achieve better than 50 pm resolution and magnifications of up to about 10,000,000x whereas most light microscopes are limited by diffraction to about 200 nm resolution and useful magnifications below 2000x.

Transmission electron microscopes use electrostatic and electromagnetic lenses to control the electron beam and focus it to form an image. These electron optical lenses are analogous to the glass lenses of an optical light microscope.

Electron microscopes are used to investigate the ultrastructure of a wide range of biological and inorganic specimens including microorganisms, cells, large molecules, biopsy samples, metals, and crystals. Industrially, electron microscopes are often used for quality control and failure analysis. Modern electron microscopes produce electron micrographs using specialized digital cameras and frame grabbers to capture the image.

History

The first electromagnetic lens was developed in 1926 by Hans Busch.

According to Dennis Gabor, the physicist Leó Szilárd tried in 1928 to convince Busch to build an electron microscope, for which he had filed a patent.

German physicist Ernst Ruska and the electrical engineer Max Knoll constructed the prototype electron microscope in 1931, capable of four-hundred-power magnification; the apparatus was the first demonstration of the principles of electron microscopy. Two years later, in 1933, Ruska built an electron microscope that exceeded the resolution attainable with an optical (light) microscope. Moreover, Reinhold Rudenberg, the scientific director of Siemens-Schuckertwerke, obtained the patent for the electron microscope in May 1931.

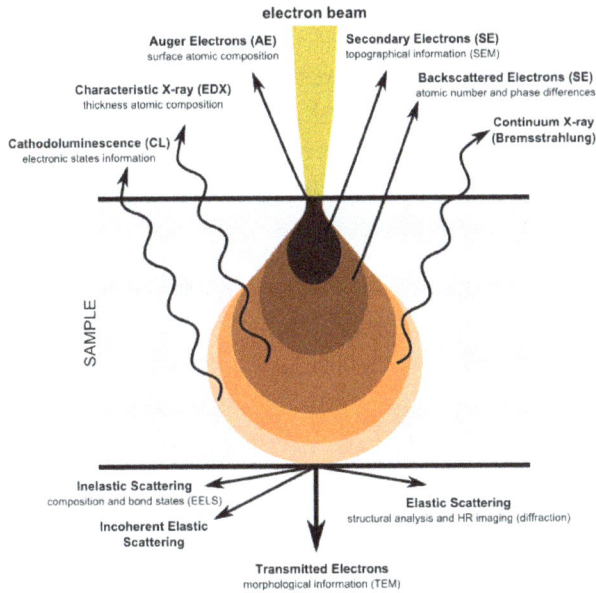

Diagram illustrating the phenomena resulting of the interaction of highly energetic electrons with matter.

In 1932, Ernst Lubcke of Siemens & Halske built and obtained images from a prototype electron microscope, applying concepts described in the Rudenberg patent applications. Five years later (1937), the firm financed the work of Ernst Ruska and Bodo von Borries, and employed Helmut Ruska (Ernst's brother) to develop applications for the microscope, especially with biological specimens. Also in 1937, Manfred von Ardenne pioneered the scanning electron microscope. The first commercial electron microscope was produced in 1938 by Siemens. The first North American electron microscope was constructed in 1938, at the University of Toronto, by Eli Franklin Burton and students Cecil Hall, James Hillier, and Albert Prebus; and Siemens produced a transmission electron microscope (TEM) in 1939. Although contemporary transmission electron microscopes are capable of two million-power magnification, as scientific instruments, they remain based upon Ruska's prototype.

Types

Transmission Electron Microscope (TEM)

The original form of electron microscope, the transmission electron microscope (TEM) uses a high voltageelectron beam to illuminate the specimen and create an image. The electron beam is produced by an electron gun, commonly fitted with a tungsten filament cathode as the electron source. The electron beam is accelerated by an anode typically at +100 keV (40 to 400 keV) with respect to the cathode, focused by electrostatic and electromagnetic lenses, and transmitted through the specimen that is in part transparent to electrons and in part scatters them out of the beam. When it emerges from the specimen, the electron beam carries information about the structure of the specimen that is magnified by the objective lens system of the microscope. The spatial variation in this information (the "image") may be viewed by projecting the magnified electron image onto a fluorescent viewing screen coated with a phosphor or scintillator material such as zinc sulfide. Alternatively, the image can be photographically recorded by exposing a photographic film or plate directly to the electron beam, or a high-resolution phosphor may be coupled by means of a lens optical system or a fibre optic light-guide to the sensor of a digital camera. The image detected by the digital camera may be displayed on a monitor or computer.

The resolution of TEMs is limited primarily by spherical aberration, but a new generation of aberration correctors have been able to partially overcome spherical aberration to increase resolution. Hardware correction of spherical aberration for the high-resolution transmission electron microscopy (HRTEM) has allowed the production of images with resolution below 0.5 angstrom (50 picometres) and magnifications above 50 million times. The ability to determine the positions of atoms within materials has made the HRTEM an important tool for nano-technologies research and development.

Transmission electron microscopes are often used in electron diffraction mode. The advantages of electron diffraction over X-ray crystallography are that the specimen need not be a single crystal or even a polycrystalline powder, and also that the Fourier transform reconstruction of the object's magnified structure occurs physically and thus avoids the need for solving the phase problem faced by the X-ray crystallographers after obtaining their X-ray diffraction patterns of a single crystal or polycrystalline powder.

The major disadvantage of the transmission electron microscope is the need for extremely thin sections of the specimens, typically about 100 nanometers. Biological specimens are typically required to be chemically fixed, dehydrated and embedded in a polymer resin to stabilize them sufficiently to allow ultrathin sectioning. Sections of biological specimens, organic polymers and similar materials may require special treatment with heavy atom labels in order to achieve the required image contrast.

Scanning Electron Microscope (SEM)

The SEM produces images by probing the specimen with a focused electron beam that is scanned across a rectangular area of the specimen (raster scanning). When the electron beam interacts with the specimen, it loses energy by a variety of mechanisms. The lost energy is converted into alternative forms such as heat, emission of low-energy secondary electrons and high-energy backscattered electrons, light emission (cathodoluminescence) or X-ray emission, all of which provide signals carrying information about the properties of the specimen surface, such as its topography and composition. The image displayed by an SEM maps the varying intensity of any of these signals into the image in a position corresponding to the position of the beam on the specimen when the signal was generated. In the SEM image of an ant shown below and to the right, the image was constructed from signals produced by a secondary electron detector, the normal or conventional imaging mode in most SEMs.

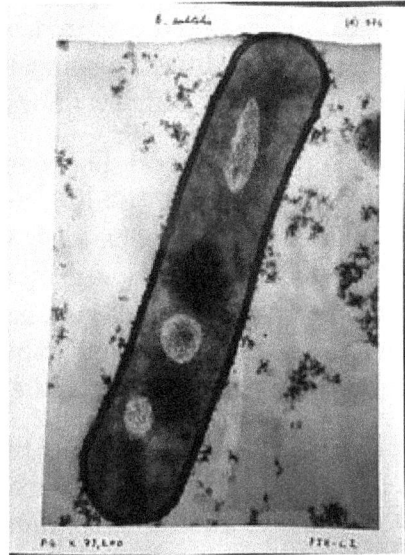

Image of bacillus subtilis taken with a 1960s electron microscope.

Generally, the image resolution of an SEM is at least an order of magnitude poorer than that of a TEM. However, because the SEM image relies on surface processes rather than transmission, it is able to image bulk samples up to many centimeters in size and (depending on instrument design and settings) has a great depth of field, and so can produce images that are good representations of the three-dimensional shape of the sample. Another advantage of SEM is its variety called environmental scanning electron microscope (ESEM) can produce images of sufficient quality and resolution with the samples being wet or contained in low vacuum or gas. This greatly facilitates imaging biological samples that are unstable in the high vacuum of conventional electron microscopes.

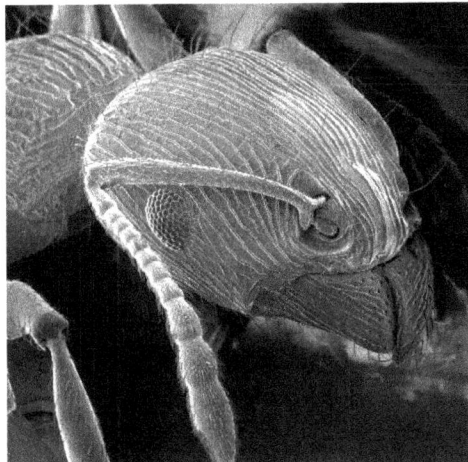

An image of an ant in a scanning electron microscope

Color

In their most common configurations, electron microscopes produce images with a single brightness value per pixel, with the results usually rendered in grayscale. However, often these images are then colorized through the use of feature-detection software, or simply by hand-editing using a graphics editor. This may be done to clarify structure or for aesthetic effect and generally does not add new information about the specimen.

In some configurations information about several specimen properties is gathered per pixel, usually by the use of multiple detectors. In SEM, the attributes of topography and material contrast can be obtained by a pair of backscattered electron detectors and such attributes can be superimposed in a single color image by assigning a different primary color to each attribute. Similarly, a combination of backscattered and secondary electron signals can be assigned to different colors and superimposed on a single color micrograph displaying simultaneously the properties of the specimen.

Some types of detectors used in SEM have analytical capabilities, and can provide several items of data at each pixel. Examples are the Energy-dispersive X-ray spectroscopy (EDS) detectors used in elemental analysis and Cathodoluminescence microscope (CL) systems that analyse the intensity and spectrum of electron-induced luminescence in (for example) geological specimens. In SEM systems using these detectors it is common to color code the signals and superimpose them in a single color image, so that differences in the distribution of the various components of the specimen can be seen clearly and compared. Optionally, the standard secondary electron image can be merged with the one or more compositional channels, so that the specimen's structure and composition can be compared. Such images can be made while maintaining the full integrity of the original signal, which is not modified in any way.

Reflection Electron Microscope (REM)

In the reflection electron microscope (REM) as in the TEM, an electron beam is incident on a surface but instead of using the transmission (TEM) or secondary electrons (SEM), the reflected beam of elastically scattered electrons is detected. This technique is typically coupled with reflection high energy electron diffraction (RHEED) and *reflection high-energy loss spectroscopy (RHELS)*. Another variation is spin-polarized low-energy electron microscopy (SPLEEM), which is used for looking at the microstructure of magnetic domains.

Scanning Transmission Electron Microscope (STEM)

The STEM rasters a focused incident probe across a specimen that (as with the TEM) has been thinned to facilitate detection of electrons scattered *through* the specimen. The high resolution of the TEM is thus possible in STEM. The focusing action (and aberrations) occur before the electrons hit the specimen in the STEM, but afterward in the TEM. The STEMs use of SEM-like beam rastering simplifies annular dark-field imaging, and other analytical techniques, but also means that image data is acquired in serial rather than in parallel fashion. Often TEM can be equipped with the scanning option and then it can function both as TEM and STEM.

Sample Preparation

An insect coated in gold for viewing with a scanning electron microscope

Materials to be viewed under an electron microscope may require processing to produce a suitable sample. The technique required varies depending on the specimen and the analysis required:

- *Chemical fixation* – for biological specimens aims to stabilize the specimen's mobile macromolecular structure by chemical crosslinking of proteins with aldehydes such as formaldehyde and glutaraldehyde, and lipids with osmium tetroxide.

- *Negative stain* – suspensions containing nanoparticles or fine biological material (such as viruses and bacteria) are briefly mixed with a dilute solution of an electron-opaque solution such as ammonium molybdate, uranyl acetate (or formate), or phosphotungstic acid. This mixture is applied to a suitably coated EM grid, blotted, then allowed to dry. Viewing of this preparation in the TEM should be carried out without delay for best results. The method is important in microbiology for fast but crude morphological identification, but can also be used as the basis for high resolution 3D reconstruction using EM tomography methodology when carbon films are used for support. Negative staining is also used for observation of nanoparticles.

- *Cryofixation* – freezing a specimen so rapidly, in liquid ethane, and maintained at liquid nitrogen or even liquid helium temperatures, so that the water forms vitreous (non-crystalline) ice. This preserves the specimen in a snapshot of its solution state. An entire field called cryo-electron microscopy has branched from this technique. With the development of cryo-electron microscopy of vitreous sections (CEMOVIS), it is now possible to observe samples from virtually any biological specimen close to its native state.

- *Dehydration* – or replacement of water with organic solvents such as ethanol or acetone, followed by critical point drying or infiltration with embedding resins. Also freeze drying.

- *Embedding, biological specimens* – after dehydration, tissue for observation in the transmission electron microscope is embedded so it can be sectioned ready for viewing. To do this the tissue is passed through a 'transition solvent' such as propylene oxide (epoxypropane) or acetone and then infiltrated with an epoxyresin such as Araldite, Epon, or Durcupan; tissues may also be embedded directly in water-miscible acrylic resin. After the resin has been polymerized (hardened) the sample is thin sectioned (ultrathin sections) and stained – it is then ready for viewing.

- *Embedding, materials* – after embedding in resin, the specimen is usually ground and polished to a mirror-like finish using ultra-fine abrasives. The polishing process must be performed carefully to minimize scratches and other polishing artifacts that reduce image quality.

- *Metal shadowing* – Metal (e.g. platinum) is evaporated from an overhead electrode and applied to the surface of a biological sample at an angle. The surface topography results in variations in the thickness of the metal that are seen as variations in brightness and contrast in the electron microscope image.

- *Replication* – A surface shadowed with metal (e.g. platinum, or a mixture of carbon and platinum) at an angle is coated with pure carbon evaporated from carbon electrodes at right angles to the surface. This is followed by removal of the specimen material (e.g. in an acid bath, using enzymes or by mechanical separation) to produce a surface replica that records the surface ultrastructure and can be examined using transmission electron microscopy.

- *Sectioning* – produces thin slices of specimen, semitransparent to electrons. These can be cut on an ultramicrotome with a diamond knife to produce ultra-thin sections about 60–90 nm thick. Disposable glass knives are also used because they can be made in the lab and are much cheaper.

- *Staining* – uses heavy metals such as lead, uranium or tungsten to scatter imaging electrons and thus give contrast between different structures, since many (especially biological) materials are nearly "transparent" to electrons (weak phase objects). In biology, specimens can be stained "en bloc" before embedding and also later after sectioning. Typically thin sections are stained for several minutes with an aqueous or alcoholic solution of uranyl acetate followed by aqueous lead citrate.

- *Freeze-fracture or freeze-etch* – a preparation method particularly useful for examining lipid membranes and their incorporated proteins in "face on" view. The fresh tissue or cell suspension is frozen rapidly (cryofixation), then fractured by breaking or by using a microtome while maintained at liquid nitrogen temperature. The cold fractured surface (sometimes "etched" by increasing the temperature to about −100 °C for several minutes to let some ice sublime) is then shadowed with evaporated platinum or gold at an average angle of 45° in a high vacuum evaporator. A second coat of carbon, evaporated perpendicular to the average surface plane is often performed to improve stability of the replica coating. The specimen is returned to room temperature and pressure, then the extremely fragile "pre-shadowed" metal replica of the fracture surface is released from the underlying biological material by careful chemical digestion with acids, hypochlorite solution or SDS detergent. The still-floating replica is thoroughly washed free from residual chemicals, carefully fished up on fine grids, dried then viewed in the TEM.

- *Ion beam milling* – thins samples until they are transparent to electrons by firing ions (typically argon) at the surface from an angle and sputtering material from the surface. A subclass of this is focused ion beam milling, where gallium ions are used to produce an electron transparent membrane in a specific region of the sample, for example through a device within a microprocessor. Ion beam milling may also be used for cross-section polishing prior to SEM analysis of materials that are difficult to prepare using mechanical polishing.

- *Conductive coating* – an ultrathin coating of electrically conducting material, deposited either by high vacuum evaporation or by low vacuum sputter coating of the sample. This is done to prevent the accumulation of static electric fields at the specimen due to the electron irradiation required during imaging. The coating materials include gold, gold/palladium, platinum, tungsten, graphite, etc.

- *Earthing* – to avoid electrical charge accumulation on a conductive coated sample, it is usually electrically connected to the metal sample holder. Often an electrically conductive adhesive is used for this purpose.

Disadvantages

Electron microscopes are expensive to build and maintain, but the capital and running costs of confocal light microscope systems now overlaps with those of basic electron microscopes. Microscopes designed to achieve high resolutions must be housed in stable buildings (sometimes underground) with special services such as magnetic field cancelling systems.

The samples largely have to be viewed in vacuum, as the molecules that make up air would scatter the electrons. An exception is the environmental scanning electron microscope, which allows hydrated samples to be viewed in a low-pressure (up to 20 Torr or 2.7 kPa) and/or wet environment.

Scanning electron microscopes operating in conventional high-vacuum mode usually image conductive specimens; therefore non-conductive materials require conductive coating (gold/palladium alloy, carbon, osmium, etc.). The low-voltage mode of modern microscopes makes possible the observation of non-conductive specimens without coating. Non-conductive materials can be imaged also by a variable pressure (or environmental) scanning electron microscope.

Small, stable specimens such as carbon nanotubes, diatom frustules and small mineral crystals (asbestos fibres, for example) require no special treatment before being examined in the electron microscope. Samples of hydrated materials, including almost all biological specimens have to be prepared in various ways to stabilize them, reduce their thickness (ultrathin sectioning) and increase their electron optical contrast (staining). These processes may result in *artifacts*, but these can usually be identified by comparing the results obtained by using radically different specimen preparation methods. It is generally believed by scientists working in the field that as results from various preparation techniques have been compared and that there is no reason that they should all produce similar artifacts, it is reasonable to believe that electron microscopy features correspond with those of living cells. Since the 1980s, analysis of cryofixed, vitrified specimens has also become increasingly used by scientists, further confirming the validity of this technique.

Applications

Semiconductor and data storage	Materials research
• Circuit edit	• Device testing and characterization
• Defect analysis	• Electron beam-induced deposition
• Failure analysis	• Materials qualification
Biology and life sciences	• Medical research
• 3D tissue imaging	• Nanometrology
• Biological production and viral load monitoring	• Nanoprototyping
• Cryobiology	Industry
• Cryo-electron microscopy	• 2D & 3D micro-characterization
• Diagnostic electron microscopy	• Chemical/Petrochemical
	• Direct beam-writing fabrication

• Drug research	• Dynamic materials experiments
• Electron tomography	• Forensics
• Food science	• Fractography and failure analysis
• Particle analysis	• High-resolution imaging
• Protein localization	• Macro sample to nanometer metrology
• Structural biology	• Mining (mineral liberation analysis)
• Toxicology	• Particle detection and characterization
• Virology	• Pharmaceutical QC
• Vitrification	• Sample preparation

References

• Bannister, Niel (November 1, 2012). Langton, Phil, ed. Essential Guide to Reading Biomedical Papers: Recognizing and Interpreting Best Practice. Wiley-Blackwell. doi:10.1002/9781118402184. ISBN 9781118402184.

• Molleman, Areles (March 6, 2003). Patch Clamping: An Introductory Guide To Patch Clamp Electrophysiology. Wiley. doi:10.1002/0470856521. ISBN 9780470856529. Retrieved November 2, 2014.

• Linley, John (2013). "11". In Gamper, Nikita. Perforated Whole-Cell Patch-Clamp Recording(PDF) (Second ed.). Humana Press. pp. 149–157. ISBN 978-1-62703-351-0. Retrieved November 10, 2014.

• Sally Satel; Scott O. Lilienfeld (2015). Brainwashed: The Seductive Appeal of Mindless Neuroscience. Basic Books. ISBN 978-0465062911.

• Truesdell, C.A. (1980). The Tragicomical History of Thermodynamics, 1822–1854, Springer, New York, ISBN 0-387-90403-4.

• Lebon, G., Jou, D., Casas-Vázquez, J. (2008). Understanding Non-equilibrium Thermodynamics: Foundations, Applications, Frontiers, Springer-Verlag, Berlin, ISBN 978-3-540-74252-4.

• McMurry, John (2011). Organic chemistry: with biological applications (2nd ed.). Belmont, CA: Brooks/Cole. p. 395. ISBN 9780495391470.

• Hostettmann, K; Marston, A; Hostettmann, M (1998). Preparative Chromatography Techniques Applications in Natural Product Isolation (Second ed.). Berlin, Heidelberg: Springer Berlin Heidelberg. p. 50. ISBN 9783662036310.

• Harwood, Laurence M.; Moody, Christopher J. (1989). Experimental organic chemistry: Principles and Practice (Illustrated ed.). WileyBlackwell. pp. 180–185. ISBN 978-0-632-02017-1.

• Anfinsen, Christian B.; Edsall, John Tileston; Richards, Frederic Middlebrook, eds. (1976). Advances in Protein Chemistry. pp. 6–7. ISBN 978-0-12-034230-3.

• Bailon, Pascal; Ehrlich, George K.; Fung, Wen-Jian and Berthold, Wolfgang (2000) An Overview of Affinity Chromatography, Humana Press. ISBN 978-0-89603-694-9.

• Edward I. Solomon; A. B. P. Lever (3 February 2006). Inorganic electronic structure and spectroscopy. Wiley-Interscience. p. 78. ISBN 978-0-471-97124-5. Retrieved 29 April 2011.

3

Allied Fields of Biophysics

The allied fields of biophysics are medical biophysics, clinical biophysics, membrane biophysics, molecular biophysics and biophysical chemistry. Medical physics is the application of physics and the methods that are used in medicine. This chapter is a compilation of the various branches of biophysics that form an integral part of the subject matter.

Medical Physics

Medical physics (also called biomedical physics, medical biophysics or applied physics in medicine) is, generally speaking, the application of physics concepts, theories and methods to medicine or healthcare. Medical physics departments may be found in hospitals or universities.

In the case of hospital work, the term 'Medical Physicist' is the title of a specific healthcare profession with a specific mission statement. Such Medical Physicists are often found in the following healthcare specialties: diagnostic and intervention radiology (also known as medical imaging), nuclear medicine, and radiation oncology (also known as radiotherapy). However, areas of specialty are widely varied in scope and breadth, e.g. clinical physiology (also known as physiological measurement, several countries), neurophysiology (Finland), radiation protection (many countries), and audiology (Netherlands).

University departments are of two types. The first type are mainly concerned with preparing students for a career as a hospital medical physicist and research focuses on improving the practice of the profession. A second type (increasingly called 'biomedical physics') has a much wider scope and may include research in any applications of physics to medicine from the study of biomolecular structure to microscopy and nanomedicine. For example, physicist Richard Feynman theorized about the future of nanomedicine. He wrote about the idea of a *medical* use for biological machines. Feynman and Albert Hibbs suggested that certain repair machines might one day be reduced in size to the point that it would be possible to (as Feynman put it) "swallow the doctor". The idea was discussed in Feynman's 1959 essay *There's Plenty of Room at the Bottom*.

Mission Statement of Medical Physicists

In the case of hospital medical physics departments, the mission statement is the following:

"Medical Physicists will contribute to maintaining and improving the quality, safety and cost-effectiveness of healthcare services through patient-oriented activities requiring expert action, involvement or advice regarding the specification, selection, acceptance testing, commissioning, quality assurance/control and optimised clinical use of medical devices and regarding patient risks and protection from associated physical agents (e.g., x-rays, electromagnetic fields, laser light, radionuclides) including the prevention of unintended or accidental exposures; all activities will be based on current best evidence or own scientific research when the available evidence is not sufficient. The scope includes risks to volunteers in biomedical research, carers and comforters. The scope often includes risks to workers and public particularly when these impact patient risk"

The term "physical agents" refers to ionising and non-ionising electromagnetic radiations, static electric and magnetic fields, ultrasound, laser light and any other Physical Agent associated with medical e.g., x-rays in computerised tomography (CT), gamma rays/radionuclides in nuclear medicine, magnetic fields and radio-frequencies in magnetic resonance imaging (MRI), ultrasound in ultrasound imaging and Doppler measurements etc.

This mission includes the following 11 key activities:

1. Scientific problem solving service: Comprehensive problem solving service involving recognition of less than optimal performance or optimised use of medical devices, identification and elimination of possible causes or misuse, and confirmation that proposed solutions have restored device performance and use to acceptable status. All activities are to be based on current best scientific evidence or own research when the available evidence is not sufficient.

2. Dosimetry measurements: Measurement of doses suffered by patients, volunteers in biomedical research, carers, comforters and persons subjected to non-medical imaging exposures (e.g., for legal or employment purposes); selection, calibration and maintenance of dosimetry related instrumentation; independent checking of dose related quantities provided by dose reporting devices (including software devices); measurement of dose related quantities required as inputs to dose reporting or estimating devices (including software). Measurements to be based on current recommended techniques and protocols. Includes dosimetry of all physical agents.

3. Patient safety/risk management (including volunteers in biomedical research, carers, comforters and persons subjected to non-medical imaging exposures. Surveillance of medical devices and evaluation of clinical protocols to ensure the ongoing protection of patients, volunteers in biomedical research, carers, comforters and persons subjected to non-medical imaging exposures from the deleterious effects of physical agents in accordance with the latest published evidence or own research when the available evidence is not sufficient. Includes the development of risk assessment protocols.

4. Occupational and public safety/risk management (when there is an impact on medical exposure or own safety). Surveillance of medical devices and evaluation of clinical protocols with respect to protection of workers and public when impacting the exposure of patients, volunteers in biomedical research, carers, comforters and persons subjected to non-medical imaging exposures or responsibility with respect to own safety. Includes the development of risk assessment protocols in conjunction with other experts involved in occupational / public risks.

5. Clinical medical device management: Specification, selection, acceptance testing, commissioning and quality assurance/ control of medical devices in accordance with the latest published European or International recommendations and the management and supervision of associated programmes. Testing to be based on current recommended techniques and protocols.

6. Clinical involvement: Carrying out, participating in and supervising everyday radiation protection and quality control procedures to ensure ongoing effective and optimised use of medical radiological devices and including patient specific optimization.

7. Development of service quality and cost-effectiveness: Leading the introduction of new medical radiological devices into clinical service, the introduction of new medical physics services and participating in the introduction/development of clinical protocols/techniques whilst giving due attention to economic issues.

8. Expert consultancy: Provision of expert advice to outside clients (e.g., clinics with no in-house medical physics expertise).

9. Education of healthcare professionals (including medical physics trainees: Contributing to quality healthcare professional education through knowledge transfer activities concerning the technical-scientific knowledge, skills and competences supporting the clinically effective, safe, evidence-based and economical use of medical radiological devices. Participation in the education of medical physics students and organisation of medical physics residency programmes.

10. Health technology assessment (HTA): Taking responsibility for the physics component of health technology assessments related to medical radiological devices and /or the medical uses of radioactive substances/sources.

11. Innovation: Developing new or modifying existing devices (including software) and protocols for the solution of hitherto unresolved clinical problems.

Medical Biophysics and Biomedical Physics

Some education institutions house departments or programs bearing the title "medical biophysics" or "biomedical physics" or "applied physics in medicine". Generally, these fall into one of two categories: interdisciplinary departments that house biophysics, radiobiology, and medical physics under a single umbrella; and undergraduate programs that prepare students for further study in medical physics, biophysics, or medicine.

Areas of Speciality

The International Organization for Medical Physics (IOMP) - the world's premier professional organization for medical physics with nearly 22,000 members in 84 countries - recognizes main areas of medical physics employment and focus. These are:

Medical Imaging Physics

Para-sagittal MRI of the head in a patient with benign familial macrocephaly.

Medical imaging physics is also known as diagnostic and interventional radiology physics. Clinical (both "in-house" and "consulting") physicists typically deal with areas of testing, optimization, and quality assurance of diagnostic radiology physics areas such as radiographic X-rays, fluoroscopy, mammography, angiography, and computed tomography, as well as non-ionizing radiation modalities such as ultrasound, and MRI. They may also be engaged with radiation protection issues such as radiation exposure monitoring and dosimetry. In addition, many imaging physicists are often also involved with nuclear medicine systems, including single photon emission computed tomography (SPECT) and positron emission tomography (PET). Sometimes, imaging physicists may be engaged in clinical areas, but for research and teaching purposes, such as e.g. quantifying intravascular ultrasound as a possible method of imaging a particular vascular object.

Radiation Therapeutic Physics

Radiation therapeutic physics is also known as radiotherapy physics or radiation oncology physics. The majority of medical physicists currently working in the US, Canada, and some western countries are of this group. A Radiation Therapy physicist typically deals with linear accelerator (Linac) systems and kilovoltage x-ray treatment units on a daily basis, as well as more advanced modalities such as TomoTherapy, gamma knife, cyberknife, proton therapy, and brachytherapy. The academic and research side of therapeutic physics may encompass fields such as boron neutron capture therapy, sealed source radiotherapy, terahertz radiation, high-intensity focused ultrasound (including lithotripsy), optical radiationlasers, ultraviolet etc. including photodynamic therapy, as well as nuclear medicine including unsealed source radiotherapy, and photomedicine, which is the use of light to treat and diagnose disease.

Nuclear Medicine Physics

This is a branch of medicine that uses radiation to provide information about the functioning of a person's specific organs or to treat disease. In most cases, the information is used by physicians to make a quick, accurate diagnosis of the patient's illness. The thyroid, bones, heart, liver and many other organs can be easily imaged, and disorders in their function revealed. In some cases radiation can be used to treat diseased organs, or tumours. Five Nobel Laureates have been intimately involved with the use of radioactive tracers in medicine. Over 10,000 hospitals worldwide use radioisotopes in medicine, and about 90% of the procedures are for diagnosis. The most common radioisotope used in diagnosis is technetium-99, with some 30 million procedures per year, accounting for 80% of all nuclear medicine procedures worldwide. In developed countries (26% of world population) the frequency of diagnostic nuclear medicine is 1.9% per year, and the frequency of therapy with radioisotopes is about one tenth of this. In the USA there are some 18 million nuclear medicine procedures per year among 311 million people, and in Europe about 10 million among 500 million people. In Australia there are about 560,000 per year among 21 million people, 470,000 of these using reactor isotopes. The use of radiopharmaceuticals in diagnosis is growing at over 10% per year. Nuclear medicine was developed in the 1950s by physicians with an endocrine emphasis, initially using iodine-131 to diagnose and then treat thyroid disease. In recent years, specialists have also come from radiology, as dual CT/PET procedures have become established. Computed X-ray tomography (CT) scans and nuclear medicine contribute 36% of the total radiation exposure and 75% of the medical exposure to the US population, according to a US National Council on Radiation Protection & Measurements report in 2009. The report showed that Americans' average total yearly radiation exposure had increased from 3.6 millisievert to 6.2 mSv per year since the early 1980s, due to medical-related procedures. (Industrial radiation exposure, including that from nuclear power plants, is less than 0.1% of overall public radiation exposure.)

Health Physics

Health physics is also known as Physical Hazards or Physical Safety or Radiation Safety or Radiation Protection.

- Background radiation

- Radiation protection

- Dosimetry

- Health physics

- Radiological protection of patients

Clinical Audiology Physics

- Physics and acoustics for audiology: the basic physics of sound, instrumentation, and the principles of digital signal processing involved in audiological research. Topics include the physics of sound waves, room acoustics, the measurement of reverberation time; the nature of acoustic impedance; the nature of filters and amplifiers, acoustics of speech, calibration.

Laser Medicine

- Lasers and applications in medicine

Medical Optics

- Optical imaging

Neurophysics

Neurophysics (or neural physics) is also known as functional neuroimaging. It refers to imaging structure and function in the nervous system as well as:

- fMRI
- molecular imaging
- electrical impedance tomography
- diffuse optical imaging
- optical coherence tomography
- dual energy X-ray absorptiometry
- image-guided surgery

Cardiophysics

- Cardiophysics (or cardiovascular physics) is using the methods of, and theories from, physics to study cardiovascular system at different levels of its organisation, from the molecular scale to whole organisms.
- Electrocardiography

Physiological Measurement Techniques

Physiological measurements have also been used to monitor and measure various physiological parameters. Many physiological measurement techniques are non-invasive and can be used in conjunction with, or as an alternative to, other invasive methods.

- Electroencephalography
- Electromyography

- Electronystagmography
- Endoscopy
- Medical ultrasonography
- Non-ionising radiation (lasers, ultraviolet etc.)
- Near infrared spectroscopy
- Pulse oximetry
- Blood gas monitor
- Blood pressure measurement

Physics of the Human and Animal Bodies

(8.1) *Biomechanics*

- Body mass index
- Sports biomechanics
- Biofluid mechanics

(8.2) *Bioelectromagnetics* of the human and animal bodies

- Bioelectrodynamics
- Biomagnetism
- Electromagnetic radiation and health
- Electrophysiology

(8.3) *Biomaterials and artificial organs*

- Biomaterials for cell and organ therapies
- Cardiovascular biomaterials, artificial heart and cardiac assist devices
- Artificial skin, bones, joints, teeth and related biomaterials
- Histopathology
- Drug delivery and control release

Healthcare Informatics and Computational Physics

- Information and communication in medicine

- Medical informatics

- Image processing, display and visualization

- Computer-aided diagnosis

- Picture archiving and communication systems (PACS)

- Standards: DICOM, ISO, IHE

- Hospital information systems

- e-Health

- Telemedicine

- Digital operating room

- Workflow, patient-specific modeling

- Medicine on the Internet of Things

- Distant monitoring and telehomecare

Areas of Research and Academic Development

ECG trace

Non-clinical physicists may or may not focus on the above areas from an academic and research point of view, but their scope of specialization may also encompass lasers and ultraviolet systems (such as photodynamic therapy), fMRI and other methods for functional imaging as well as molecular imaging, electrical impedance tomography, diffuse optical imaging, optical coherence tomography, and dual energy X-ray absorptiometry.

Education and Training

In Australia and New Zealand

The Australasian College of Physical Scientists and Engineers in Medicine (ACPSEM) is the professional body that oversees the education and certification of medical physicists in Australia and New Zealand and has a mission to advance services and professional standards in medical physics and biomedical engineering.

In Europe

The presence of Medical Physicists at Expert level ('Medical Physics Experts') in healthcare in Europe is required by EC Directive 2013/59/EURATOM. The European Federation of Organizations for Medical Physics (EFOMP) has defined a detailed inventory of learning outcomes for Medical Physics Experts in terms of Knowledge, Skills and Competences (the latter in Europe means 'responsibilities'). In Europe the professional preparation for Medical Physicists consists of a first degree in Physics or equivalent (e.g., biophysics, electrical or mechanical engineering), a Masters in Medical Physics and a 2-year training Residency. For the latest EFOMP policy statement on the role and education and training requirements of the Medical Physicist and Medical Physics Expert go to. For the final version of the EC funded document 'European Guidelines on the Medical Physics Expert' download Report 174 and its annexes from.

In North America

In North America, medical physics training is offered at the master's, doctorate, post-doctorate and/or residency levels. Also, a professional doctorate has been recently introduced as an option. Several universities offer these degrees in Canada and the United States.

As of October 2013, over 70 universities in North America have medical physics graduate programs or residencies that are accredited by *The Commission on Accreditation of Medical Physics Education Programs* (CAMPEP). The majority of residencies are therapy, but diagnostic and nuclear are also on the rise in the past several years.

Professional certification is obtained from the American Board of Radiology (for all 4 areas), the American Board of Medical Physics (for MRI), the American Board of Science in Nuclear Medicine (for Nuc Med and PET), and the Canadian College of Physicists in Medicine. As of 2012, enrollment in a CAMPEP-accredited residency or graduate program is required to start the ABR certification process. Starting in 2014, completion of a CAMPEP-accredited residency will be required to advance to part 2 of the ABR certification process.

United Kingdom

From October 2011 as part of the Modernising Scientific Careers scheme, the route to accreditation as a medical physicist in England and Wales is provided by the Scientist Training Programme (STP). This scheme is a three-year graduate scheme provided by the National School of Healthcare Science. Entrants are required to have a prior under-graduate degree (1:1 or 2:1) in an appropriate physical science.

The STP involves a part-time MSc in Medical Physics (provided by either King's College London, University of Liverpool or University of Newcastle) in addition to practical training within the National Health Service. Assessment is provided by the completion of competencies and by a final assessment similar to the OSCE under-taken by other clinical staff. Completion of the STP leads to accreditation with the Institute of Physics and Engineering in Medicine (IPEM) and registration as a Clinical Scientist.

Prior to 2011 the training route in the United Kingdom was administered in two parts, and this scheme is still used in Scotland and Northern Ireland). Part I involved limited clinical experience and a full-time MSc in medical physics. Part II involved exclusive-ly clinical experience in which the candidate would produce a portfolio of experience which would be submitted to the Academy for Healthcare Science which (in addition to a viva) would lead to professional accreditation with IPEM.

International

There are regular regional and international educational medical physics activities. The oldest of these is the International College on Medical Physics at the International Cen-tre for Theoretical Physics (ICTP), Trieste, Italy. This College has educated more than 1000 medical physicists from developing countries.

Legislative and Advisory Bodies

- ICRU: International Commission on Radiation Units and Measurements

- ICRP: International Commission on Radiological Protection

- NCRP: National Council on Radiation Protection & Measurements

- NRC: Nuclear Regulatory Commission

- FDA: Food and Drug Administration

- IAEA: International Atomic Energy Agency

Clinical Biophysics

Clinical biophysics is that branch of medical science that studies the action process and the effects of non-ionising physical energies utilised for therapeutic purposes. Physical energy can be applied for diagnostic or therapeutic aims.

The principle on which clinical biophysics is based are represented by the recognizability and the specificity of the physical signal applied:

- recognizability: the capacity of the biological target to recognise the presence of the physical energy: this aspect becomes more important with the lowering of the energy applied.

- specificity: the capacity of the physical agent applied to the biological target to obtain a response depending on its physical characteristics: frequency, length, energy, etc. The effects do not necessarily depend on the quantity of energy applied to the biological target.

Definition

Several papers show that the response of a biological system when exposed to non-ionizing physical stimuli is not necessarily dependent on the amount of energy applied. Specific combinations of amplitude, frequency and waveform may trigger the most intense response. For example, cell proliferation or activation of metabolic pathways. This has been demonstrated for:

a) mechanical strains directly applied to the cells or tissue;

b) mechanical energy applied by ultrasound;

c) electromagnetic field exposure;

d) electric field exposure.

Several pre-clinical experiences have laid the foundation to identify exposure conditions that may be used in humans to treat diseases or to promote tissue healing. The identification of the best parameters to apply in any particular circumstance is the current goal of research activities in the field.

Physical Stimuli

Medical Applications

- Orthopaedics
- PEMF
- LIPUS
- CCEF
- Direct current

- Neurology

- Plastic surgery

- Oncology

Membrane Biophysics

Membrane biophysics is the study of biological membrane structure and function using physical, computational, mathematical, and biophysical methods. A combination of these methods can be used to create phase diagrams of different types of membranes, which yields information on thermodynamic behavior of a membrane and its components. As opposed to membrane biology, membrane biophysics focuses on quantitative information and modeling of various membrane phenomena, such as lipid raft formation, rates of lipid and cholesterol flip-flop, protein-lipid coupling, and the effect of bending and elasticity functions of membranes on inter-cell connections.

Molecular Biophysics

Molecularbiophysics is a rapidly evolving interdisciplinary area of research that combines concepts in physics, chemistry, engineering, mathematics and biology. It seeks to understand biomolecular systems and explain biological function in terms of molecular structure, structural organization, and dynamic behaviour at various levels of complexity (from single molecules to supramolecular structures, viruses and small living systems). The technical challenges are formidable, and the discipline has required development of specialized equipment and procedures capable of imaging and manipulating minute living structures, as well as novel experimental approaches.

Spectroscopy in Molecular Biophysics

Spectroscopic techniques like FT-NMR, spin labelelectron spin resonance, laser Raman, FT-infrared, circular dichroism, etc. have been widely used to understand structural dynamics of important bio-molecules and inter-molecular interactions.

Training

The field overlaps with both research into other areas of biophysics and biochemistry, teaching at the undergraduate, graduate, and professional level, and the use of the field and its related fields to develop, test, and use improved clinical therapies. Therefore, training will usually involve a graduate (usually, for teaching, clinical testing and care, and other research, a doctoral) degree and/or a degree in the health sciences, such as a medical degree. A postdoctoral training period or fellowship can also be pursued.

Biophysical Chemistry

Biophysical chemistry is a physical science that uses the concepts of physics and physical chemistry for the study of biological systems. The most common feature of the research in this subject is to seek explanation of the various phenomena in biological systems in terms of either the molecules that make up the system or the supra-molecular structure of these systems.

Techniques

Biophysical chemists employ various techniques used in physical chemistry to probe the structure of biological systems. These techniques include spectroscopic methods such as nuclear magnetic resonance (NMR) and X-ray diffraction. For example, the work for which Nobel Prize was awarded in 2009 to three chemists was based on x-ray diffraction studies of ribosomes. Some of the areas in which biophysical chemists engage themselves are protein structure and the functional structure of cell membranes. For example, enzyme action can be explained in terms of the shape of a pocket in the protein molecule that matches the shape of the substrate molecule or its modification due to binding of a metal ion. Similarly the structure and function of the biomembranes may be understood through the study of model supramolecular structures as liposomes or phospholipidvesicles of different compositions and sizes.

Institutes

The oldest reputed institute for biophysical chemistry is the Max Planck Institute for Biophysical Chemistry in Göttingen.

References

- Ashrafuzzaman, Mohammad; Tuszynski, Jack A. (2012). Membrane Biophysics. Berlin Heidelberg: Springer-Verlag. ISBN 978-3-642-16105-6.

- Peter Jomo Walla (8 July 2014). Modern Biophysical Chemistry: Detection and Analysis of Biomolecules. Wiley. pp. 1–. ISBN 978-3-527-68354-3.

- "CAMPEP Accredited Professional Doctorate Programs in Medical Physics". CAMPEP. CAMPEP. Retrieved 2015-03-29.

- CAMPEP Accredited Graduate Programs in Medical Physics. Campep.org (2011-06-01). Retrieved on 2011-06-25.

Biophysical Structure and Phenomena

The study of chemical reactions is known as enzyme kinetics. Enzymes activate these reactions and act as catalysts. Molecular motor, myoglobin, hemoglobin, pancreatic ribonuclease and antibody are some of the features explained in the following section. The section strategically encompasses and incorporates the main aspects of biophysical structure and phenomena, providing a complete understanding.

Enzyme Kinetics

Dihydrofolate reductase from *E. coli* with its two substrates dihydrofolate (right) and NADPH (left), bound in the active site. The protein is shown as a ribbon diagram, with alpha helices in red, beta sheets in yellow and loops in blue. Generated from 7DFR.

Enzyme kinetics is the study of the chemical reactions that are catalysed by enzymes. In enzyme kinetics, the reaction rate is measured and the effects of varying the conditions of the reaction are investigated. Studying an enzyme's kinetics in this way can reveal the catalytic mechanism of this enzyme, its role in metabolism, how its activity is controlled, and how a drug or an agonist might inhibit the enzyme.

Enzymes are usually proteinmolecules that manipulate other molecules — the enzymes' substrates. These target molecules bind to an enzyme's active site and are transformed into products through a series of steps known as the enzymatic mechanism

$$E + S \rightleftarrows ES \rightleftarrows ES^* \rightleftarrows EP \rightleftarrows E + P$$

These mechanisms can be divided into single-substrate and multiple-substrate mechanisms. Kinetic studies on enzymes that only bind one substrate, such as triosephosphate isomerase, aim to measure the affinity with which the enzyme binds this substrate and the turnover rate. Some other examples of enzymes are phosphofructokinase and hexokinase, both of which are important for cellular respiration (glycolysis).

When enzymes bind multiple substrates, such as dihydrofolate reductase, enzyme kinetics can also show the sequence in which these substrates bind and the sequence in which products are released. An example of enzymes that bind a single substrate and release multiple products are proteases, which cleave one protein substrate into two polypeptide products. Others join two substrates together, such as DNA polymerase linking a nucleotide to DNA. Although these mechanisms are often a complex series of steps, there is typically one *rate-determining step* that determines the overall kinetics. This rate-determining step may be a chemical reaction or a conformational change of the enzyme or substrates, such as those involved in the release of product(s) from the enzyme.

Knowledge of the enzyme's structure is helpful in interpreting kinetic data. For example, the structure can suggest how substrates and products bind during catalysis; what changes occur during the reaction; and even the role of particular amino acid residues in the mechanism. Some enzymes change shape significantly during the mechanism; in such cases, it is helpful to determine the enzyme structure with and without bound substrate analogues that do not undergo the enzymatic reaction.

Not all biological catalysts are protein enzymes; RNA-based catalysts such as ribozymes and ribosomes are essential to many cellular functions, such as RNA splicing and translation. The main difference between ribozymes and enzymes is that RNA catalysts are composed of nucleotides, whereas enzymes are composed of amino acids. Ribozymes also perform a more limited set of reactions, although their reaction mechanisms and kinetics can be analysed and classified by the same methods.

General Principles

The reaction catalysed by an enzyme uses exactly the same reactants and produces exactly the same products as the uncatalysed reaction. Like other catalysts, enzymes do not alter the position of equilibrium between substrates and products. However, unlike uncatalysed chemical reactions, enzyme-catalysed reactions display saturation kinetics. For a given enzyme concentration and for relatively low substrate concentrations, the reaction rate increases linearly with substrate concentration; the enzyme molecules are largely free to catalyse the reaction, and increasing substrate concentration means an increasing rate at which the enzyme and substrate molecules encounter one another. However, at relatively high substrate concentrations, the reaction rate asymptotically approaches the theoretical maximum; the enzyme active sites are almost all occupied and the reaction rate is determined by the intrinsic turnover rate of the enzyme. The substrate concentration midway between these two limiting cases is denoted by K_M.

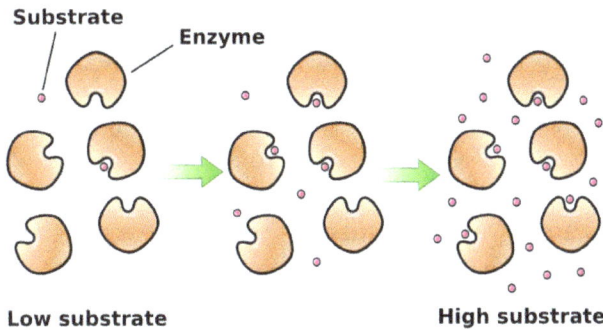

As larger amounts of substrate are added to a reaction, the available enzyme binding sites become filled to the limit of V_{max}. Beyond this limit the enzyme is saturated with substrate and the reaction rate ceases to increase.

The two most important kinetic properties of an enzyme are how quickly the enzyme becomes saturated with a particular substrate, and the maximum rate it can achieve. Knowing these properties suggests what an enzyme might do in the cell and can show how the enzyme will respond to changes in these conditions.

Enzyme Assays

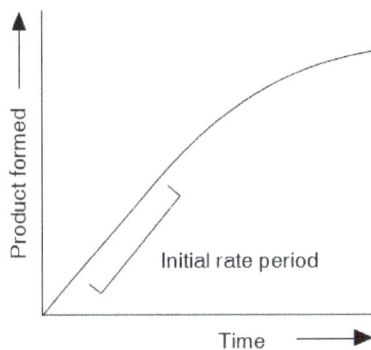

Progress curve for an enzyme reaction. The slope in the initial rate period is the initial rate of reactionv. The Michaelis–Menten equation describes how this slope varies with the concentration of substrate.

Enzyme assays are laboratory procedures that measure the rate of enzyme reactions. Because enzymes are not consumed by the reactions they catalyse, enzyme assays usually follow changes in the concentration of either substrates or products to measure the rate of reaction. There are many methods of measurement. Spectrophotometric assays observe change in the absorbance of light between products and reactants; radiometric assays involve the incorporation or release of radioactivity to measure the amount of product made over time. Spectrophotometric assays are most convenient since they allow the rate of the reaction to be measured continuously. Although radiometric assays require the removal and counting of samples (i.e., they are discontinuous assays) they are usually extremely sensitive and can measure very low levels of enzyme activity. An analogous approach is to use mass spectrometry to monitor the incorporation or release of stable isotopes as substrate is converted into product.

The most sensitive enzyme assays use lasers focused through a microscope to observe changes in single enzyme molecules as they catalyse their reactions. These measurements either use changes in the fluorescence of cofactors during an enzyme's reaction mechanism, or of fluorescent dyes added onto specific sites of the protein to report movements that occur during catalysis. These studies are providing a new view of the kinetics and dynamics of single enzymes, as opposed to traditional enzyme kinetics, which observes the average behaviour of populations of millions of enzyme molecules.

An example progress curve for an enzyme assay is shown above. The enzyme produces product at an initial rate that is approximately linear for a short period after the start of the reaction. As the reaction proceeds and substrate is consumed, the rate continuously slows (so long as substrate is not still at saturating levels). To measure the initial (and maximal) rate, enzyme assays are typically carried out while the reaction has progressed only a few percent towards total completion. The length of the initial rate period depends on the assay conditions and can range from milliseconds to hours. However, equipment for rapidly mixing liquids allows fast kinetic measurements on initial rates of less than one second. These very rapid assays are essential for measuring pre-steady-state kinetics, which are discussed below.

Most enzyme kinetics studies concentrate on this initial, approximately linear part of enzyme reactions. However, it is also possible to measure the complete reaction curve and fit this data to a non-linear rate equation. This way of measuring enzyme reactions is called progress-curve analysis. This approach is useful as an alternative to rapid kinetics when the initial rate is too fast to measure accurately.

Single-substrate Reactions

Enzymes with single-substrate mechanisms include isomerases such as triosephosphateisomerase or bisphosphoglycerate mutase, intramolecular lyases such as adenylate cyclase and the hammerhead ribozyme, an RNA lyase. However, some enzymes that only have a single substrate do not fall into this category of mechanisms. Catalase is an example of this, as the enzyme reacts with a first molecule of hydrogen peroxide substrate, becomes oxidised and is then reduced by a second molecule of substrate. Although a single substrate is involved, the existence of a modified enzyme intermediate means that the mechanism of catalase is actually a ping–pong mechanism, a type of mechanism that is discussed in the *Multi-substrate reactions* section below.

Michaelis–Menten Kinetics

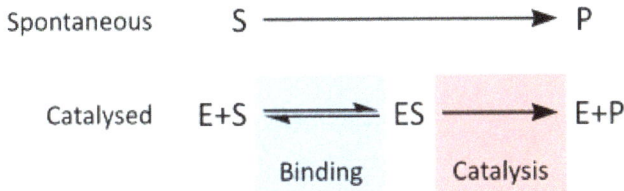

Spontaneous   S $\longrightarrow$ P

Catalysed   E+S $\rightleftharpoons$ ES $\longrightarrow$ E+P

Binding   Catalysis

A chemical reaction mechanism with or without enzyme catalysis.
The enzyme (E) binds substrate (S) to produce product (P).

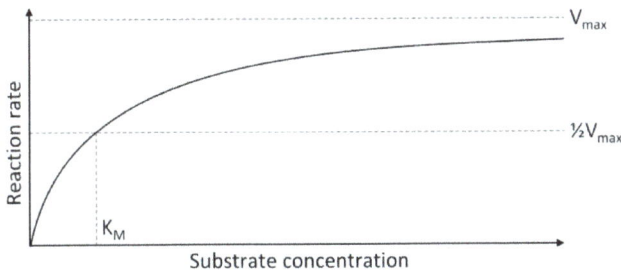

Saturation curve for an enzyme reaction showing the relation
between the substrate concentration and reaction rate.

As enzyme-catalysed reactions are saturable, their rate of catalysis does not show a linear response to increasing substrate. If the initial rate of the reaction is measured over a range of substrate concentrations (denoted as [S]), the reaction rate (v) increases as [S] increases, as shown on the right. However, as [S] gets higher, the enzyme becomes saturated with substrate and the rate reaches V_{max}, the enzyme's maximum rate.

The Michaelis–Menten kinetic model of a single-substrate reaction is shown on the right. There is an initial bimolecular reaction between the enzyme E and substrate S to form the enzyme–substrate complex ES. The rate of enzymatic reaction increases with the increase of the substrate concentration up to a certain level called V_{max}; at V_{max}, increase in substrate concentration does not cause any increase in reaction rate as there no more enzyme (E) available for reacting with substrate (S). Here, the rate of reaction becomes dependent on the ES complex and the reaction becomes a unimolecular reaction with an order of zero. Though the enzymatic mechanism for the unimolecular reaction $ES->[k_{cat}]E+P$ can be quite complex, there is typically one rate-determining enzymatic step that allows this reaction to be modelled as a single catalytic step with an apparent unimolecular rate constant k_{cat}. If the reaction path proceeds over one or several intermediates, k_{cat} will be a function of several elementary rate constants, whereas in the simplest case of a single elementary reaction (e.g. no intermediates) it will be identical to the elementary unimolecular rate constant k_2. The apparent unimolecular rate constant k_{cat} is also called *turnover number* and denotes the maximum number of enzymatic reactions catalysed per second.

The Michaelis–Menten equation describes how the (initial) reaction rate v_0 depends on the position of the substrate-binding equilibrium and the rate constant k_2.

$$v_0 = \frac{V_{max}[S]}{K_M + [S]} \quad \text{(Michaelis–Menten equation)}$$

with the constants

$$K_M \overset{\text{def}}{=} \frac{k_2 + k_{-1}}{k_1} \approx K_D$$

$$V_{max} \overset{\text{def}}{=} k_{cat}[E]_{tot}$$

This Michaelis–Menten equation is the basis for most single-substrate enzyme kinetics. Two crucial assumptions underlie this equation (apart from the general assumption about the mechanism only involving no intermediate or product inhibition, and there is no allostericity or cooperativity). The first assumption is the so-called quasi-steady-state assumption (or pseudo-steady-state hypothesis), namely that the concentration of the substrate-bound enzyme (and hence also the unbound enzyme) changes much more slowly than those of the product and substrate and thus the change over time of the complex can be set to zero $d[ES]/dt = 0$. The second assumption is that the total enzyme concentration does not change over time, thus $[E]_{tot} = [E] + [ES] = \text{const.}$. A complete derivation can be found here.

The Michaelis constant K_M is experimentally defined as the concentration at which the rate of the enzyme reaction is half V_{max}, which can be verified by substituting $[S] = K_m$ into the Michaelis–Menten equation and can also be seen graphically. If the rate-determining enzymatic step is slow compared to substrate dissociation ($k_2 \ll k_{-1}$), the Michaelis constant K_M is roughly the dissociation constant K_D of the ES complex.

If $[S]$ is small compared to K_M then the term $[S]/(K_M + [S]) \approx [S]/K_M$ and also very little ES complex is formed, thus $[E]_0 \approx [E]$. Therefore, the rate of product formation is

$$v_0 \approx \frac{k_{cat}}{K_M}[E][S] \qquad \text{if } [S] \ll K_M$$

Thus the product formation rate depends on the enzyme concentration as well as on the substrate concentration, the equation resembles a bimolecular reaction with a corresponding pseudo-second order rate constant k_2/K_M. This constant is a measure of catalytic efficiency. The most efficient enzymes reach a k_2/K_M in the range of $10^8 - 10^{10}$ M^{-1} s^{-1}. These enzymes are so efficient they effectively catalyse a reaction each time they encounter a substrate molecule and have thus reached an upper theoretical

limit for efficiency (diffusion limit); and are sometimes referred to as kinetically perfect enzymes.

Direct use of the Michaelis–Menten Equation for Time Course Kinetic Analysis

The observed velocities predicted by the Michaelis–Menten equation can be used to
directly model the time course disappearance of substrate and the production of product through incorporation of the Michaelis–Menten equation into the equation for first
order chemical kinetics. This can only be achieved however if one recognises the problem associated with the use of Euler's number in the description of first order chemical
kinetics. i.e. e^{-k} is a split constant that introduces a systematic error into calculations
and can be rewritten as a single constant which represents the remaining substrate
after each time period.

$$[S] = [S]_0 (1-k)^t$$

$$[S] = [S]_0 (1-v/[S]_0)^t$$

$$[S] = [S]_0 (1-(V_{max}[S]_0/(K_M+[S]_0)/[S]_0))^t$$

In 1983 Stuart Beal (and also independently Santiago Schnell and Claudio Mendoza
in 1997) derived a closed form solution for the time course kinetics analysis of the Michaelis-Menten mechanism. The solution, known as the Schnell-Mendoza equation,
has the form:

$$\frac{[S]}{K_M} = W[F(t)]$$

where W[] is the Lambert-W function. and where F(t) is

$$F(t) = \frac{[S]_0}{K_M} \exp\left(\frac{[S]_0}{K_M} - \frac{V_{max}}{K_M}t\right)$$

This equation is encompassed by the equation below, obtained by Berberan-Santos
(MATCH Commun. Math. Comput. Chem. 63 (2010) 283), which is also valid when the
initial substrate concentration is close to that of enzyme,

$$\frac{[S]}{K_M} = W[F(t)] - \frac{V_{max}}{k_{cat}K_M} \frac{W[F(t)]}{1+W[F(t)]}$$

where W[] is again the Lambert-W function.

Linear Plots of the Michaelis–Menten Equation

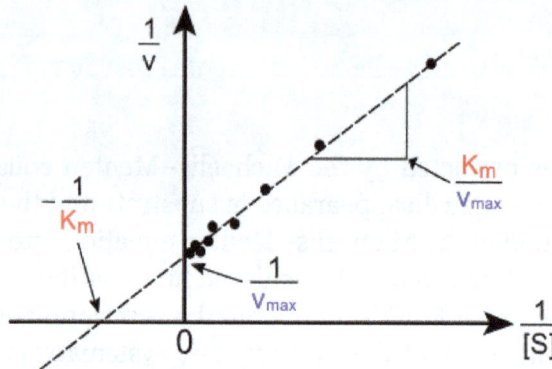

Lineweaver–Burk or double-reciprocal plot of kinetic data, showing the significance of the axis intercepts and gradient.

The plot of v versus [S] above is not linear; although initially linear at low [S], it bends over to saturate at high [S]. Before the modern era of nonlinear curve-fitting on computers, this nonlinearity could make it difficult to estimate K_M and V_{max} accurately. Therefore, several researchers developed linearisations of the Michaelis–Menten equation, such as the Lineweaver–Burk plot, the Eadie–Hofstee diagram and the Hanes–Woolf plot. All of these linear representations can be useful for visualising data, but none should be used to determine kinetic parameters, as computer software is readily available that allows for more accurate determination by nonlinear regression methods.

The Lineweaver–Burk plot or double reciprocal plot is a common way of illustrating kinetic data. This is produced by taking the reciprocal of both sides of the Michaelis–Menten equation. As shown on the right, this is a linear form of the Michaelis–Menten equation and produces a straight line with the equation $y = mx + c$ with a y-intercept equivalent to $1/V_{max}$ and an x-intercept of the graph representing $-1/K_M$.

$$\frac{1}{v} = \frac{K_M}{V_{max}[S]} + \frac{1}{V_{max}}$$

Naturally, no experimental values can be taken at negative $1/[S]$; the lower limiting value $1/[S] = 0$ (the y-intercept) corresponds to an infinite substrate concentration, where $1/v = 1/V_{max}$ as shown at the right; thus, the x-intercept is an extrapolation of the experimental data taken at positive concentrations. More generally, the Lineweaver–Burk plot skews the importance of measurements taken at low substrate concentrations and, thus, can yield inaccurate estimates of V_{max} and K_M. A more accurate linear plotting method is the Eadie-Hofstee plot. In this case, v is plotted against $v/[S]$. In the third common linear representation, the Hanes-Woolf plot, $[S]/v$ is plotted against $[S]$. In general, data normalisation can help diminish the amount of experimental work and can increase the reliability of the output, and is suitable for both graphical and numerical analysis.

Practical Significance of Kinetic Constants

The study of enzyme kinetics is important for two basic reasons. Firstly, it helps explain how enzymes work, and secondly, it helps predict how enzymes behave in living organisms. The kinetic constants defined above, K_M and V_{max}, are critical to attempts to understand how enzymes work together to control metabolism.

Making these predictions is not trivial, even for simple systems. For example, oxaloacetate is formed by malate dehydrogenase within the mitochondrion. Oxaloacetate can then be consumed by citrate synthase, phosphoenolpyruvate carboxykinase or aspartate aminotransferase, feeding into the citric acid cycle, gluconeogenesis or aspartic acid biosynthesis, respectively. Being able to predict how much oxaloacetate goes into which pathway requires knowledge of the concentration of oxaloacetate as well as the concentration and kinetics of each of these enzymes. This aim of predicting the behaviour of metabolic pathways reaches its most complex expression in the synthesis of huge amounts of kinetic and gene expression data into mathematical models of entire organisms. Alternatively, one useful simplification of the metabolic modelling problem is to ignore the underlying enzyme kinetics and only rely on information about the reaction network's stoichiometry, a technique called flux balance analysis.

Michaelis–Menten Kinetics with Intermediate

One could also consider the less simple case

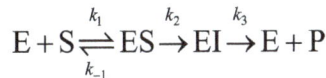

$$E + S \underset{k_{-1}}{\overset{k_1}{\rightleftharpoons}} ES \overset{k_2}{\rightarrow} EI \overset{k_3}{\rightarrow} E + P$$

where a complex with the enzyme and an intermediate exists and the intermediate is converted into product in a second step. In this case we have a very similar equation

$$v_0 = k_{cat} \frac{[S][E]_0}{K'_M + [S]}$$

but the constants are different

$$K'_M \overset{def}{=} \frac{k_3}{k_2 + k_3} K_M = \frac{k_3}{k_2 + k_3} \cdot \frac{k_2 + k_{-1}}{k_1}$$

$$k_{cat} \overset{def}{=} \frac{k_3 k_2}{k_2 + k_3}$$

We see that for the limiting case $k_3 \gg k_2$, thus when the last step from $EI -> E + P$ is much faster than the previous step, we get again the original equation. Mathematically we have then $K'_M \approx K_M$ and $k_{cat} \approx k_2$.

Multi-substrate Reactions

Multi-substrate reactions follow complex rate equations that describe how the substrates bind and in what sequence. The analysis of these reactions is much simpler if the concentration of substrate A is kept constant and substrate B varied. Under these conditions, the enzyme behaves just like a single-substrate enzyme and a plot of v by [S] gives apparent K_M and V_{max} constants for substrate B. If a set of these measurements is performed at different fixed concentrations of A, these data can be used to work out what the mechanism of the reaction is. For an enzyme that takes two substrates A and B and turns them into two products P and Q, there are two types of mechanism: ternary complex and ping–pong.

Ternary-complex Mechanisms

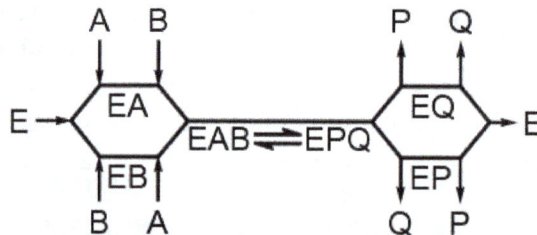

Random-order ternary-complex mechanism for an enzyme reaction. The reaction path is shown as a line and enzyme intermediates containing substrates A and B or products P and Q are written below the line.

In these enzymes, both substrates bind to the enzyme at the same time to produce an EAB ternary complex. The order of binding can either be random (in a random mechanism) or substrates have to bind in a particular sequence (in an ordered mechanism). When a set of v by [S] curves (fixed A, varying B) from an enzyme with a ternary-complex mechanism are plotted in a Lineweaver–Burk plot, the set of lines produced will intersect.

Enzymes with ternary-complex mechanisms include glutathione S-transferase,dihydrofolate reductase and DNA polymerase. The following links show short animations of the ternary-complex mechanisms of the enzymes dihydrofolate reductase[β] and DNA polymerase[γ].

Ping–pong Mechanisms

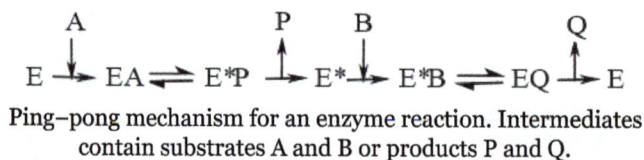

Ping–pong mechanism for an enzyme reaction. Intermediates contain substrates A and B or products P and Q.

As shown on the right, enzymes with a ping-pong mechanism can exist in two states, E and a chemically modified form of the enzyme E*; this modified enzyme is known as an intermediate. In such mechanisms, substrate A binds, changes the enzyme to E* by, for example, transferring a chemical group to the active site, and is then released. Only after the first substrate is released can substrate B bind and react with the modified enzyme, regenerating the unmodified E form. When a set of v by [S] curves (fixed A, varying B) from an enzyme with a ping–pong mechanism are plotted in a Lineweaver–Burk plot, a set of parallel lines will be produced. This is called a secondary plot.

Enzymes with ping–pong mechanisms include some oxidoreductases such as thioredoxin peroxidase,transferases such as acylneuraminate cytidylyltransferase and serine proteases such as trypsin and chymotrypsin. Serine proteases are a very common and diverse family of enzymes, including digestive enzymes (trypsin, chymotrypsin, and elastase), several enzymes of the blood clotting cascade and many others. In these serine proteases, the E* intermediate is an acyl-enzyme species formed by the attack of an active site serine residue on a peptide bond in a protein substrate. A short animation showing the mechanism of chymotrypsin is linked here.[8]

Non-Michaelis–Menten Kinetics

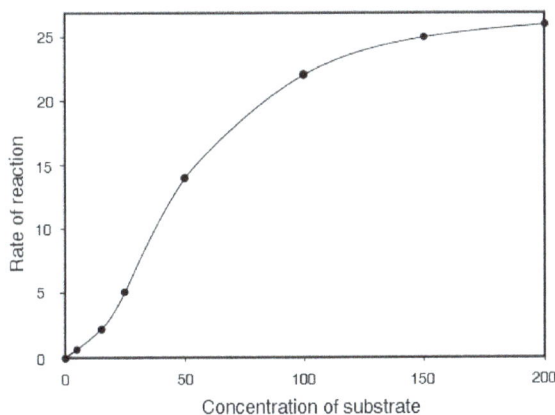

Saturation curve for an enzyme reaction showing sigmoid kinetics.

Some enzymes produce a sigmoid$v$ by [S] plot, which often indicates cooperative binding of substrate to the active site. This means that the binding of one substrate molecule affects the binding of subsequent substrate molecules. This behavior is most common in multimeric enzymes with several interacting active sites. Here, the mechanism of cooperation is similar to that of hemoglobin, with binding of substrate to one active site altering the affinity of the other active sites for substrate molecules. Positive cooperativity occurs when binding of the first substrate molecule *increases* the affinity of the other active sites for substrate. Negative cooperativity occurs when binding of the first substrate *decreases* the affinity of the enzyme for other substrate molecules.

Allosteric enzymes include mammalian tyrosyl tRNA-synthetase, which shows negative cooperativity, and bacterial aspartate transcarbamoylase and phosphofructokinase, which show positive cooperativity.

Cooperativity is surprisingly common and can help regulate the responses of enzymes to changes in the concentrations of their substrates. Positive cooperativity makes enzymes much more sensitive to [S] and their activities can show large changes over a narrow range of substrate concentration. Conversely, negative cooperativity makes enzymes insensitive to small changes in [S].

The Hill equation (biochemistry) is often used to describe the degree of cooperativity quantitatively in non-Michaelis–Menten kinetics. The derived Hill coefficient n measures how much the binding of substrate to one active site affects the binding of substrate to the other active sites. A Hill coefficient of <1 indicates negative cooperativity and a coefficient of >1 indicates positive cooperativity.

Pre-steady-state Kinetics

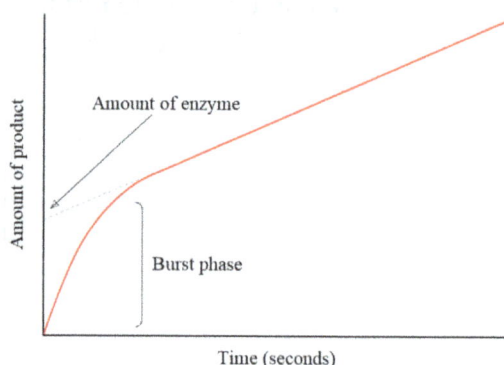

Pre-steady state progress curve, showing the burst phase of an enzyme reaction.

In the first moment after an enzyme is mixed with substrate, no product has been formed and no intermediates exist. The study of the next few milliseconds of the reaction is called pre-steady-state kinetics. Pre-steady-state kinetics is therefore concerned with the formation and consumption of enzyme–substrate intermediates (such as ES or E*) until their steady-state concentrations are reached.

This approach was first applied to the hydrolysis reaction catalysed by chymotrypsin. Often, the detection of an intermediate is a vital piece of evidence in investigations of what mechanism an enzyme follows. For example, in the ping–pong mechanisms that are shown above, rapid kinetic measurements can follow the release of product P and measure the formation of the modified enzyme intermediate E*. In the case of chymotrypsin, this intermediate is formed by an attack on the substrate by the nucleophilic serine in the active site and the formation of the acyl-enzyme intermediate.

In the figure to the right, the enzyme produces E* rapidly in the first few seconds of the reaction. The rate then slows as steady state is reached. This rapid burst phase of the reaction measures a single turnover of the enzyme. Consequently, the amount of product released in this burst, shown as the intercept on the y-axis of the graph, also gives the amount of functional enzyme which is present in the assay.

Chemical Mechanism

An important goal of measuring enzyme kinetics is to determine the chemical mechanism of an enzyme reaction, i.e., the sequence of chemical steps that transform substrate into product. The kinetic approaches discussed above will show at what rates intermediates are formed and inter-converted, but they cannot identify exactly what these intermediates are.

Kinetic measurements taken under various solution conditions or on slightly modified enzymes or substrates often shed light on this chemical mechanism, as they reveal the rate-determining step or intermediates in the reaction. For example, the breaking of a covalent bond to a hydrogenatom is a common rate-determining step. Which of the possible hydrogen transfers is rate determining can be shown by measuring the kinetic effects of substituting each hydrogen by deuterium, its stable isotope. The rate will change when the critical hydrogen is replaced, due to a primary kinetic isotope effect, which occurs because bonds to deuterium are harder to break than bonds to hydrogen. It is also possible to measure similar effects with other isotope substitutions, such as $^{13}C/^{12}C$ and $^{18}O/^{16}O$, but these effects are more subtle.

Isotopes can also be used to reveal the fate of various parts of the substrate molecules in the final products. For example, it is sometimes difficult to discern the origin of an oxygen atom in the final product; since it may have come from water or from part of the substrate. This may be determined by systematically substituting oxygen's stable isotope ^{18}O into the various molecules that participate in the reaction and checking for the isotope in the product. The chemical mechanism can also be elucidated by examining the kinetics and isotope effects under different pH conditions, by altering the metal ions or other bound cofactors, by site-directed mutagenesis of conserved amino acid residues, or by studying the behaviour of the enzyme in the presence of analogues of the substrate(s).

Enzyme Inhibition and Activation

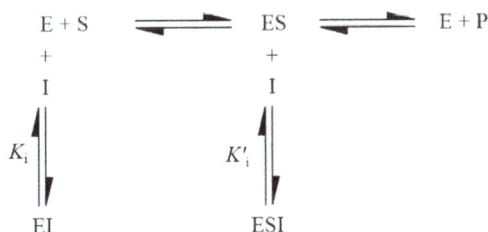

Kinetic scheme for reversible enzyme inhibitors.

Enzyme inhibitors are molecules that reduce or abolish enzyme activity, while enzyme activators are molecules that increase the catalytic rate of enzymes. These interactions can be either *reversible* (i.e., removal of the inhibitor restores enzyme activity) or *irreversible* (i.e., the inhibitor permanently inactivates the enzyme).

Reversible Inhibitors

Traditionally reversible enzyme inhibitors have been classified as competitive, uncompetitive, or non-competitive, according to their effects on K_m and V_{max}. These different effects result from the inhibitor binding to the enzyme E, to the enzyme–substrate complex ES, or to both, respectively. The division of these classes arises from a problem in their derivation and results in the need to use two different binding constants for one binding event. The binding of an inhibitor and its effect on the enzymatic activity are two distinctly different things, another problem the traditional equations fail to acknowledge. In noncompetitive inhibition the binding of the inhibitor results in 100% inhibition of the enzyme only, and fails to consider the possibility of anything in between. The common form of the inhibitory term also obscures the relationship between the inhibitor binding to the enzyme and its relationship to any other binding term be it the Michaelis–Menten equation or a dose response curve associated with ligand receptor binding. To demonstrate the relationship the following rearrangement can be made:

$$\frac{V_{max}}{1+\dfrac{[I]}{K_i}} = \frac{V_{max}}{\dfrac{[I]+K_i}{K_i}}$$

Adding zero to the bottom ([I]-[I])

$$\frac{\dfrac{V_{max}}{[I]+K_i}}{[I]+K_i-[I]}$$

Dividing by [I]+K$_i$

$$\frac{\dfrac{V_{max}}{1}}{1-\dfrac{[I]}{[I]+K_i}} = V_{max} - V_{max}\frac{[I]}{[I]+K_i}$$

This notation demonstrates that similar to the Michaelis–Menten equation, where the rate of reaction depends on the percent of the enzyme population interacting with substrate

fraction of the enzyme population bound by substrate

$$\frac{[S]}{[S]+K_m}$$

fraction of the enzyme population bound by inhibitor

$$\frac{[I]}{[I]+K_i}$$

the effect of the inhibitor is a result of the percent of the enzyme population interacting with inhibitor. The only problem with this equation in its present form is that it assumes absolute inhibition of the enzyme with inhibitor binding, when in fact there can be a wide range of effects anywhere from 100% inhibition of substrate turn over to just >0%. To account for this the equation can be easily modified to allow for different degrees of inhibition by including a delta V_{max} term.

$$V_{max} - \Delta V_{max} \frac{[I]}{[I]+K_i}$$

or

$$V_{max1} - (V_{max1} - V_{max2}) \frac{[I]}{[I]+K_i}$$

This term can then define the residual enzymatic activity present when the inhibitor is interacting with individual enzymes in the population. However the inclusion of this term has the added value of allowing for the possibility of activation if the secondary V_{max} term turns out to be higher than the initial term. To account for the possibly of activation as well the notation can then be rewritten replacing the inhibitor "I" with a modifier term denoted here as "X".

$$V_{max1} - (V_{max1} - V_{max2}) \frac{[X]}{[X]+K_x}$$

While this terminology results in a simplified way of dealing with kinetic effects relating to the maximum velocity of the Michaelis–Menten equation, it highlights potential problems with the term used to describe effects relating to the K_m. The K_m relating to the affinity of the enzyme for the substrate should in most cases relate to potential changes in the binding site of the enzyme which would directly result from enzyme inhibitor interactions. As such a term similar to the one proposed above to modulate V_{max} should be appropriate in most situations:

$$K_{m1} - (K_{m1} - K_{m2}) \frac{[X]}{[X]+K_x}$$

Irreversible Inhibitors

Enzyme inhibitors can also irreversibly inactivate enzymes, usually by covalently modifying active site residues. These reactions, which may be called suicide substrates, follow exponential decay functions and are usually saturable. Below saturation, they follow first order kinetics with respect to inhibitor.

Mechanisms of Catalysis

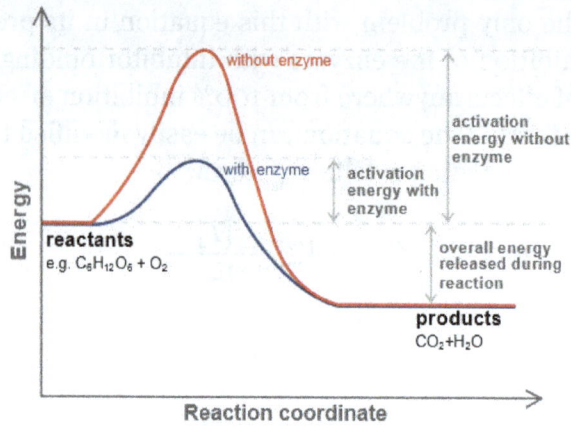

The energy variation as a function of reaction coordinate shows the stabilisation of the transition state by an enzyme.

The favoured model for the enzyme–substrate interaction is the induced fit model. This model proposes that the initial interaction between enzyme and substrate is relatively weak, but that these weak interactions rapidly induce conformational changes in the enzyme that strengthen binding. These conformational changes also bring catalytic residues in the active site close to the chemical bonds in the substrate that will be altered in the reaction. Conformational changes can be measured using circular dichroism or dual polarisation interferometry. After binding takes place, one or more mechanisms of catalysis lower the energy of the reaction's transition state by providing an alternative chemical pathway for the reaction. Mechanisms of catalysis include catalysis by bond strain; by proximity and orientation; by active-site proton donors or acceptors; covalent catalysis and quantum tunnelling.

Enzyme kinetics cannot prove which modes of catalysis are used by an enzyme. However, some kinetic data can suggest possibilities to be examined by other techniques. For example, a ping–pong mechanism with burst-phase pre-steady-state kinetics would suggest covalent catalysis might be important in this enzyme's mechanism. Alternatively, the observation of a strong pH effect on V_{max} but not K_m might indicate that a residue in the active site needs to be in a particular ionisation state for catalysis to occur.

History

In 1902 Victor Henri proposed a quantitative theory of enzyme kinetics, but at the time the experimental significance of the hydrogen ion concentration was not yet recognized. After Peter Lauritz Sørensen had defined the logarithmic pH-scale and introduced the concept of buffering in 1909 the German chemist Leonor Michaelis and his Canadian postdoc Maud Leonora Menten repeated Henri's experiments and confirmed his equation, which is now generally referred to as Michaelis-Menten kinetics (sometimes also *Henri-Michaelis-Menten kinetics*). Their work was further developed by G. E. Briggs and J. B. S. Haldane, who derived kinetic equations that are still widely considered today a starting point in modeling enzymatic activity.

The major contribution of the Henri-Michaelis-Menten approach was to think of enzyme reactions in two stages. In the first, the substrate binds reversibly to the enzyme, forming the enzyme-substrate complex. This is sometimes called the Michaelis complex. The enzyme then catalyzes the chemical step in the reaction and releases the product. The kinetics of many enzymes is adequately described by the simple Michaelis-Menten model, but all enzymes have internal motions that are not accounted for in the model and can have significant contributions to the overall reaction kinetics. This can be modeled by introducing several Michaelis-Menten pathways that are connected with fluctuating rates, which is a mathematical extension of the basic Michaelis Menten mechanism.

Software

ENZO

ENZO (Enzyme Kinetics) is a graphical interface tool for building kinetic models of enzyme catalyzed reactions. ENZO automatically generates the corresponding differential equations from a stipulated enzyme reaction scheme. These differential equations are processed by a numerical solver and a regression algorithm which fits the coefficients of differential equations to experimentally observed time course curves. ENZO allows rapid evaluation of rival reaction schemes and can be used for routine tests in enzyme kinetics.

Molecular Motor

Molecular motors are biological molecular machines that are the essential agents of movement in living organisms. In general terms, a motor is a device that consumes energy in one form and converts it into motion or mechanical work; for example, many protein-based molecular motors harness the chemical free energy released by the hydrolysis of ATP in order to perform mechanical work. In terms of energetic efficiency, this type of motor can be superior to currently available man-made motors. One important difference between molecular motors and macroscopic motors is that molecular motors operate in the thermal bath, an environment in which the fluctuations due to thermal noise are significant.

Examples

Some examples of biologically important molecular motors:

- Cytoskeletal motors

  o Myosins are responsible for muscle contraction, intracellular cargo transport, and producing cellular tension.

  o Kinesin moves cargo inside cells away from the nucleus along microtubules.

  o Dynein produces the axonemal beating of cilia and flagella and also transports cargo along microtubules towards the cell nucleus.

- Polymerisation motors

  o Actin polymerization generates forces and can be used for propulsion. ATP is used.

  o Microtubule polymerization using GTP.

  o Dynamin is responsible for the separation of clathrin buds from the plasma membrane. GTP is used.

- Rotary motors:

  o F_oF_1-ATP synthase family of proteins convert the chemical energy in ATP to the electrochemical potential energy of a proton gradient across a membrane or the other way around. The catalysis of the chemical reaction and the movement of protons are coupled to each other via the mechanical rotation of parts of the complex. This is involved in ATP synthesis in the mitochondria and chloroplasts as well as in pumping of protons across the vacuolar membrane.

  o The bacterial flagellum responsible for the swimming and tumbling of *E. coli* and other bacteria acts as a rigid propeller that is powered by a rotary motor. This motor is driven by the flow of protons across a membrane, possibly using a similar mechanism to that found in the F_o motor in ATP synthase.

- Nucleic acid motors:

  o RNA polymerase transcribes RNA from a DNA template.

  o DNA polymerase turns single-stranded DNA into double-stranded DNA.

  o Helicases separate double strands of nucleic acids prior to transcription or replication. ATP is used.

- o   Topoisomerases reduce supercoiling of DNA in the cell. ATP is used.

- o   RSC and SWI/SNF complexes remodel chromatin in eukaryotic cells. ATP is used.

- o   SMC proteins responsible for chromosome condensation in eukaryotic cells.

- o   Viral DNA packaging motors inject viral genomic DNA into capsids as part of their replication cycle, packing it very tightly. Several models have been put forward to explain how the protein generates the force required to drive the DNA into the capsid; for a review . An alternative proposal is that, in contrast with all other biological motors, the force is not generated directly by the protein, but by the DNA itself. In this model, ATP hydrolysis is used to drive protein conformational changes that alternatively dehydrate and rehydrate the DNA, cyclically driving it from B-DNA to A-DNA and back again. A-DNA is 23% shorter than B-DNA, and the DNA shrink/expand cycle is coupled to a protein-DNA grip/release cycle to generate the forward motion that propels DNA into the capsid.

- Synthetic molecular motors have been created by chemists that yield rotation, possibly generating torque.

Theoretical Considerations

Because the motor events are stochastic, molecular motors are often modeled with the Fokker-Planck equation or with Monte Carlo methods. These theoretical models are especially useful when treating the molecular motor as a Brownian motor.

Experimental Observation

In experimental biophysics, the activity of molecular motors is observed with many different experimental approaches, among them:

- Fluorescent methods: fluorescence resonance energy transfer (FRET), fluorescence correlation spectroscopy (FCS), total internal reflection fluorescence (TIRF).

- Magnetic tweezers can also be useful for analysis of motors that operate on long pieces of DNA.

- Neutron spin echo spectroscopy can be used to observe motion on nanosecond timescales.

- Optical tweezers are well-suited for studying molecular motors because of their low spring constants.

- Scattering techniques: single particle tracking based on dark field microscopy or interferometric scattering microscopy (iSCAT)
- Single-molecule electrophysiology can be used to measure the dynamics of individual ion channels.

Many more techniques are also used. As new technologies and methods are developed, it is expected that knowledge of naturally occurring molecular motors will be helpful in constructing synthetic nanoscale motors.

Non-biological

Recently, chemists and those involved in nanotechnology have begun to explore the possibility of creating molecular motors *de novo*. These synthetic molecular motors currently suffer many limitations that confine their use to the research laboratory. However, many of these limitations may be overcome as our understanding of chemistry and physics at the nanoscale increases. Systems like the nanocars, while not technically motors, are illustrative of recent efforts towards synthetic nanoscale motors.

Myoglobin

Myoglobin (symbol Mb or MB) is an iron- and oxygen-binding protein found in the muscle tissue of vertebrates in general and in almost all mammals. It is related to hemoglobin, which is the iron- and oxygen-binding protein in blood, specifically in the red blood cells. In humans, myoglobin is only found in the bloodstream after muscle injury. It is an abnormal finding, and can be diagnostically relevant when found in blood.

Myoglobin is the primary oxygen-carrying pigment of muscle tissues. High concentrations of myoglobin in muscle cells allow organisms to hold their breath for a longer period of time. Diving mammals such as whales and seals have muscles with particularly high abundance of myoglobin. Myoglobin is found in Type I muscle, Type II A and Type II B, but most texts consider myoglobin not to be found in smooth muscle.

Myoglobin was the first protein to have its three-dimensional structure revealed by X-ray crystallography. This achievement was reported in 1958 by John Kendrew and associates. For this discovery, John Kendrew shared the 1962 Nobel Prize in chemistry with Max Perutz. Despite being one of the most studied proteins in biology, its physiological function is not yet conclusively established: mice genetically engineered to lack myoglobin are viable, but showed a 30% reduction in volume of blood being pumped by the heart during a contraction. They adapted to this deficiency through natural reactions to inadequate oxygen supply (hypoxia) and a widening of blood vessels (vasodilation). In humans myoglobin is encoded by the *MB*gene.

As hemoglobin can take the forms of oxyhemoglobin (HbO_2), carboxyhemoglobin (HbCO), and methemoglobin (met-Hb), so can myoglobin take the forms of oxymyoglobin (MbO_2), carboxymyoglobin (MbCO), and metmyoglobin (met-Mb).

Differences from Hemoglobin

Myoglobin is similar to hemoglobin in that it is involved in the transportation of oxygen to cells. There are many distinct differences that set the protein apart from hemoglobin. For one the protein has only one binding site for oxygen on the one heme group on the protein. While myoglobin can only hold one oxygen, the affinity for that oxygen is very high compared to hemoglobin. This is probably because hemoglobin, transporting 4 oxygens to the tissues and muscles where myoglobin is mostly present. The myoglobin takes the oxygen from the hemoglobin (due to the Bohr Effect) and takes that oxygen to muscle cells for use in metabolic processes.

Meat Color

Myoglobin contains hemes, pigments responsible for the color of red meat. The color that meat takes is partly determined by the degree of oxidation of the myoglobin. In fresh meat the iron atom is the ferrous (+2) oxidation state bound to an oxygen molecule (O_2). Meat cooked well done is brown because the iron atom is now in the ferric (+3) oxidation state, having lost an electron. If meat has been exposed to nitrites, it will remain pink because the iron atom is bound to NO, nitric oxide (true of, e.g., corned beef or cured hams). Grilled meats can also take on a pink "smoke ring" that comes from the iron binding to a molecule of carbon monoxide. Raw meat packed in a carbon monoxide atmosphere also shows this same pink "smoke ring" due to the same principles. Notably, the surface of this raw meat also displays the pink color, which is usually associated in consumers' minds with fresh meat. This artificially induced pink color can persist, reportedly up to one year. Hormel and Cargill are both reported to use this meat-packing process, and meat treated this way has been in the consumer market since 2003.

Role in Disease

Myoglobin is released from damaged muscle tissue (rhabdomyolysis), which has very high concentrations of myoglobin. The released myoglobin is filtered by the kidneys but is toxic to the renal tubular epithelium and so may cause acute kidney injury. It is not the myoglobin itself that is toxic (it is a protoxin) but the ferrihemate portion that is dissociated from myoglobin in acidic environments (e.g., acidic urine, lysosomes).

Myoglobin is a sensitive marker for muscle injury, making it a potential marker for heart attack in patients with chest pain. However, elevated myoglobin has low specificity for acute myocardial infarction (AMI) and thus CK-MB, cardiac Troponin, ECG, and clinical signs should be taken into account to make the diagnosis.

Structure and Bonding

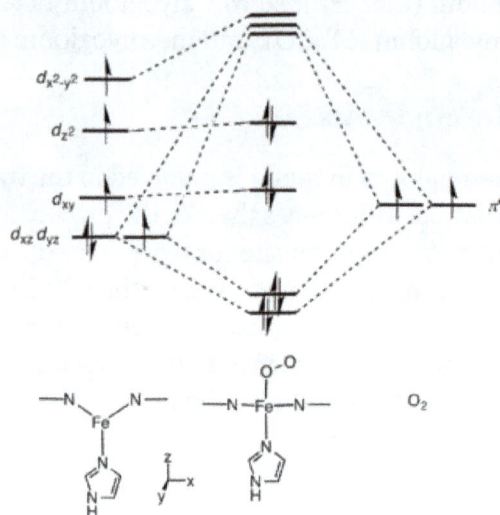

Molecular orbital description of Fe-O_2 interaction in myoglobin.

Myoglobin belongs to the globin superfamily of proteins, and as with other globins, consists of eight alpha helices connected by loops. Human globin contains 154 amino acids.

Myoglobin contains a porphyrin ring with an iron at its center. A *proximal*histidine group (His-94) is attached directly to iron, and a *distal* histidine group (His-65) hovers near the opposite face. The distal imidazole is not bonded to the iron but is available to interact with the substrate O_2. This interaction encourages the binding of O_2, but not carbon monoxide (CO), which still binds about 240× more strongly than O_2.

The binding of O_2 causes substantial structural change at the Fe center, which shrinks in radius and moves into the center of N4 pocket. O_2-binding induces "spin-pairing": the five-coordinate ferrous deoxy form is high spin and the six coordinate oxy form is low spin and diamagnetic.

Synthetic Analogues

Many models of myoglobin have been synthesized as part of a broad interest in transition metal dioxygen complexes. A well known example is the *picket fence porphyrin*, which consists of a ferrous complex of a sterically bulky derivative of tetraphenylporphyrin. In the presence of an imidazole ligand, this ferrous complex reversibly binds O_2. The O_2 substrate adopts a bent geometry, occupying the sixth position of the iron center. A key property of this model is the slow formation of the μ-oxo dimer, which is an inactive diferric state. In nature, such deactivation pathways are suppressed by protein matrix that prevents close approach of the Fe-porphyrin assemblies.

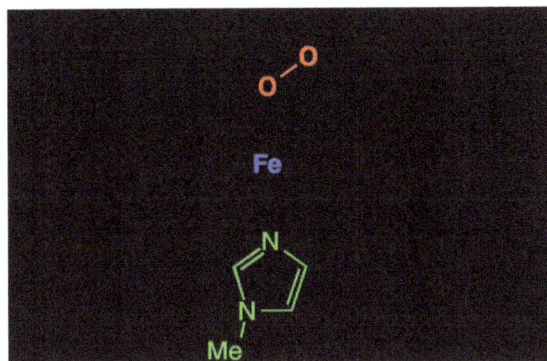

A picket-fence porphyrin complex of Fe, with axial coordination sites occupied by methylimidazole (green) and dioxygen. The R groups flank the O_2-binding site.

Hemoglobin

Hemoglobin also spelledhaemoglobin (United Kingdom spelling) and abbreviated Hb or Hgb, is the iron-containing oxygen-transport metalloprotein in the red blood cells of all vertebrates (with the exception of the fish family Channichthyidae) as well as the tissues of some invertebrates. Hemoglobin in the blood carries oxygen from the respiratory organs (lungs or gills) to the rest of the body (i.e. the tissues). There it releases the oxygen to permit aerobic respiration to provide energy to power the functions of the organism in the process called metabolism.

In mammals, the protein makes up about 96% of the red blood cells' dry content (by weight), and around 35% of the total content (including water). Hemoglobin has an oxygen-binding capacity of 1.34 mL O_2 per gram, which increases the total blood oxygen capacity seventy-fold compared to dissolved oxygen in blood. The mammalian hemoglobin molecule can bind (carry) up to four oxygen molecules.

Hemoglobin is involved in the transport of other gases: It carries some of the body's respiratory carbon dioxide (about 20–25% of the total) as carbaminohemoglobin, in which CO_2 is bound to the globin protein. The molecule also carries the important regulatory molecule nitric oxide bound to a globin protein thiol group, releasing it at the same time as oxygen.

Hemoglobin is also found outside red blood cells and their progenitor lines. Other cells that contain hemoglobin include the A9 dopaminergic neurons in the substantia nigra, macrophages, alveolar cells, and mesangial cells in the kidney. In these tissues, hemoglobin has a non-oxygen-carrying function as an antioxidant and a regulator of iron metabolism.

Hemoglobin and hemoglobin-like molecules are also found in many invertebrates, fungi, and plants. In these organisms, hemoglobins may carry oxygen, or they may act to transport and regulate other small molecules and ions such as carbon dioxide, nitric oxide, hydrogen sulfide and sulfide. A variant of the molecule, called leghemoglobin, is used to scavenge oxygen away from anaerobic systems, such as the nitrogen-fixing nodules of leguminous plants, before the oxygen can poison (deactivate) the system.

Research History

In 1825 J.F. Engelhard discovered that the ratio of Fe to protein is identical in the hemoglobins of several species. From the known atomic mass of iron he calculated the molecular mass of hemoglobin to $n \times 16000$ (n = number of iron atoms per hemoglobin, now known to be 4), the first determination of a protein's molecular mass. This "hasty conclusion" drew a lot of ridicule at the time from scientists who could not believe that any molecule could be that big. Gilbert Smithson Adair confirmed Engelhard's results in 1925 by measuring the osmotic pressure of hemoglobin solutions.

The oxygen-carrying protein hemoglobin was discovered by Hünefeld in 1840. In 1851, German physiologist Otto Funke published a series of articles in which he described growing hemoglobin crystals by successively diluting red blood cells with a solvent such as pure water, alcohol or ether, followed by slow evaporation of the solvent from the resulting protein solution. Hemoglobin's reversible oxygenation was described a few years later by Felix Hoppe-Seyler.

In 1959, Max Perutz determined the molecular structure of myoglobin (similar to hemoglobin) by X-ray crystallography. This work resulted in his sharing with John Kendrew the 1962 Nobel Prize in Chemistry.

The role of hemoglobin in the blood was elucidated by French physiologistClaude Bernard. The name *hemoglobin* is derived from the words *heme* and *globin*, reflecting the fact that each subunit of hemoglobin is a globular protein with an embedded heme group. Each heme group contains one iron atom, that can bind one oxygen molecule through ion-induced dipole forces. The most common type of hemoglobin in mammals contains four such subunits.

Genetics

Hemoglobin consists of protein subunits (the "globin" molecules), and these proteins, in turn, are folded chains of a large number of different amino acids called polypeptides. The amino acid sequence of any polypeptide created by a cell is in turn determined by the stretches of DNA called genes. In all proteins, it is the amino acid sequence that determines the protein's chemical properties and function.

There is more than one hemoglobin gene: in humans, hemoglobin A (the main form of hemoglobin present) is coded for by the genes, *HBA1*, *HBA2*, and *HBB*. The amino acid sequences of the globin proteins in hemoglobins usually differ between species. These differences grow with evolutionary distance between species. For example, the most common hemoglobin sequences in humans and chimpanzees are nearly identical, differing by only one amino acid in both the alpha and the beta globin protein chains. These differences grow larger between less closely related species.

Even within a species, different variants of hemoglobin always exist, although one sequence is usually a "most common" one in each species. Mutations in the genes for the hemoglobin protein in a species result in hemoglobin variants. Many of these mutant forms of hemoglobin cause no disease. Some of these mutant forms of hemoglobin, however, cause a group of hereditary diseases termed the *hemoglobinopathies*. The best known hemoglobinopathy is sickle-cell disease, which was the first human disease whose mechanism was understood at the molecular level. A (mostly) separate set of diseases called thalassemias involves underproduction of normal and sometimes abnormal hemoglobins, through problems and mutations in globin gene regulation. All these diseases produce anemia.

Variations in hemoglobin amino acid sequences, as with other proteins, may be adaptive. For example, hemoglobin has been found to adapt in different ways to high altitudes. Organisms living at high elevations experience lower partial pressures of oxygen compared to those at sea level. This presents a challenge to the organisms that inhabit such environments because hemoglobin, which normally binds oxygen at high partial pressures of oxygen, must be able to bind oxygen when it is present at a lower pressure. Different organisms have adapted to such a challenge. For example, recent studies have suggested genetic variants in deer mice that help explain how deer mice that live in the mountains are able to survive in the thin air that accompanies high altitudes. A researcher from the University of Nebraska-Lincoln found mutations in four different genes that can account for differences between deer mice that live in lowland prairies versus the mountains. After examining wild mice captured from both highlands and lowlands, it was found that: the genes of the two breeds are "virtually identical–except for those that govern the oxygen-carrying capacity of their hemoglobin". "The genetic difference enables highland mice to make more efficient use of their oxygen", since less is available at higher altitudes, such as those in the mountains.Mammoth hemoglobin featured mutations that allowed for oxygen delivery at lower temperatures, thus enabling mammoths to migrate to higher latitudes during the Pleistocene. This was also found in hummingbirds that inhabit the Andes. Hummingbirds already expend a lot of energy and thus have high oxygen demands and yet Andean hummingbirds have been found to thrive in high altitudes. Non-synonymous mutations in the hemoglobin gene of multiple species living at high elevations (*Oreotrochilus, A. castelnaudii, C. violifer, P. gigas,* and *A. viridicuada*) have caused the protein to have less of an affinity for inositol hexaphosphate (IHP), a molecule found in birds that has a similar role as 2,3-BPG in humans; this results in the ability to bind oxygen in lower partial pressures.

It should be noted, however, that birds' unique circulatory lungs also promote efficient use of oxygen at low partial pressures of O_2. These two adaptations reinforce each other and account for birds' remarkable high-altitude performance.

Hemoglobin adaptation extends to humans, as well. Studies have found that a small number of native Tibetan women have a genotype which codes for hemoglobin to be more highly saturated with oxygen. Natural selection seems to be the main force working on this gene because mortality rate of offspring is significantly lower for women with higher hemoglobin-oxygen affinity when compared to the mortality rate of offspring from women with low hemoglobin-oxygen affinity. While the exact genotype and mechanism by which this occurs is not yet clear, selection is acting on these women's ability to bind oxygen in low partial pressures which overall allows them to better sustain crucial metabolic processes.

Synthesis

Hemoglobin (Hb) is synthesized in a complex series of steps. The heme part is synthesized in a series of steps in the mitochondria and the cytosol of immature red blood cells, while the globin protein parts are synthesized by ribosomes in the cytosol. Production of Hb continues in the cell throughout its early development from the proerythroblast to the reticulocyte in the bone marrow. At this point, the nucleus is lost in mammalian red blood cells, but not in birds and many other species. Even after the loss of the nucleus in mammals, residual ribosomal RNA allows further synthesis of Hb until the reticulocyte loses its RNA soon after entering the vasculature (this hemoglobin-synthetic RNA in fact gives the reticulocyte its reticulated appearance and name).

Structure

Hemoglobin has a quaternary structure characteristic of many multi-subunit globular proteins. Most of the amino acids in hemoglobin form alpha helices, connected by short non-helical segments. Hydrogen bonds stabilize the helical sections inside this protein, causing attractions within the molecule, folding each polypeptide chain into a specific shape. Hemoglobin's quaternary structure comes from its four subunits in roughly a tetrahedral arrangement.

In most vertebrates, the hemoglobin molecule is an assembly of four globular protein subunits. Each subunit is composed of a protein chain tightly associated with a non-protein prostheticheme group. Each protein chain arranges into a set of alpha-helix structural segments connected together in a globin fold arrangement, so called because this arrangement is the same folding motif used in other heme/globin proteins such as myoglobin. This folding pattern contains a pocket that strongly binds the heme group.

A heme group consists of an iron (Fe) ion (charged atom) held in a heterocyclic ring, known as a porphyrin. This porphyrin ring consists of four pyrrole molecules cyclically linked together (by methine bridges) with the iron ion bound in the center. The iron ion, which is the site of oxygen binding, coordinates with the four nitrogen atoms in the center of the ring, which all lie in one plane. The iron is bound strongly (covalently) to the globular protein via the N atoms of the imidazole ring of F8 histidine residue (also known as the proximal histidine) below the porphyrin ring. A sixth position can reversibly bind oxygen by a coordinate covalent bond, completing the octahedral group of six ligands. Oxygen binds in an "end-on bent" geometry where one oxygen atom binds to Fe and the other protrudes at an angle. When oxygen is not bound, a very weakly bonded water molecule fills the site, forming a distorted octahedron.

Even though carbon dioxide is carried by hemoglobin, it does not compete with oxygen for the iron-binding positions but is bound to the protein chains of the structure.

The iron ion may be either in the Fe^{2+} or in the Fe^{3+} state, but ferrihemoglobin (methemoglobin) (Fe^{3+}) cannot bind oxygen. In binding, oxygen temporarily and reversibly oxidizes (Fe^{2+}) to (Fe^{3+}) while oxygen temporarily turns into the superoxide ion, thus iron must exist in the +2 oxidation state to bind oxygen. If superoxide ion associated to Fe^{3+} is protonated, the hemoglobin iron will remain oxidized and incapable of binding oxygen. In such cases, the enzyme methemoglobin reductase will be able to eventually reactivate methemoglobin by reducing the iron center.

In adult humans, the most common hemoglobin type is a tetramer (which contains four subunit proteins) called *hemoglobin A*, consisting of two α and two β subunits non-covalently bound, each made of 141 and 146 amino acid residues, respectively. This is denoted as $\alpha_2\beta_2$. The subunits are structurally similar and about the same size. Each subunit has a molecular weight of about 16,000 daltons, for a total molecular weight of the tetramer of about 64,000 daltons (64,458 g/mol). Thus, 1 g/dL = 0.1551 mmol/L. Hemoglobin A is the most intensively studied of the hemoglobin molecules.

In human infants, the hemoglobin molecule is made up of 2 α chains and 2 γ chains. The gamma chains are gradually replaced by β chains as the infant grows.

The four polypeptide chains are bound to each other by salt bridges, hydrogen bonds, and the hydrophobic effect.

Oxygen Saturation

In general, hemoglobin can be saturated with oxygen molecules (oxyhemoglobin), or desaturated with oxygen molecules (deoxyhemoglobin).

Oxyhemoglobin

Oxyhemoglobin is formed during physiological respiration when oxygen binds to the heme component of the protein hemoglobin in red blood cells. This process occurs in the pulmonary capillaries adjacent to the alveoli of the lungs. The oxygen then travels through the blood stream to be dropped off at cells where it is utilized as a terminal electron acceptor in the production of ATP by the process of oxidative phosphorylation. It does not, however, help to counteract a decrease in blood pH. Ventilation, or breathing, may reverse this condition by removal of carbon dioxide, thus causing a shift up in pH.

Hemoglobin exists in two forms, a *taut (tense) form* (T) and a *relaxed form* (R). Various factors such as low pH, high CO_2 and high 2,3 BPG at the level of the tissues favor the taut form, which has low oxygen affinity and releases oxygen in the tissues. Conversely, a high pH, low CO_2, or low 2,3 BPG favors the relaxed form, which can better bind oxygen. The partial pressure of the system also affects O_2 affinity where, at high partial pressures of oxygen (such as those present in the alveoli), the relaxed (high affinity, R) state is favoured. Inversely, at low partial pressures (such as those present in respiring tissues), the (low affinity, T) tense state is favoured. Additionally, the binding of oxygen to the iron(II) heme pulls the iron into the plane of the porphyrin ring, causing a slight conformational shift. The shift encourages oxygen to bind to the three remaining heme units within hemoglobin (thus, oxygen binding is cooperative).

Deoxygenated Hemoglobin

Deoxygenated hemoglobin is the form of hemoglobin without the bound oxygen. The absorption spectra of oxyhemoglobin and deoxyhemoglobin differ. The oxyhemoglobin has significantly lower absorption of the 660 nm wavelength than deoxyhemoglobin, while at 940 nm its absorption is slightly higher. This difference is used for the measurement of the amount of oxygen in a patient's blood by an instrument called a pulse oximeter. This difference also accounts for the presentation of cyanosis, the blue to purplish color that tissues develop during hypoxia.

Evolution of Vertebrate Hemoglobin

Scientists agree that the event that separated myoglobin from hemoglobin occurred after lampreys diverged from jawed vertebrates. This separation of myoglobin and hemoglobin allowed for the different functions of the two molecules to arise and develop: myoglobin has more to do with oxygen storage while hemoglobin is tasked with oxygen transport. The α- and β-like globin genes encode the individual subunits of the protein. The predecessors of these genes arose through another duplication event also after the gnathosome common ancestor derived from jawless fish, approximately 450–500 million years ago. The development of α and β genes created the potential for hemoglobin to be composed of multiple subunits, a physical composition central to hemoglobin's ability to transport oxygen. Having multiple subunits contributes to hemoglobin's ability to bind oxygen cooperatively as well as be regulated allosterically. Subsequently, the α gene also underwent a duplication event to form the *HBA1* and *HBA2* genes. These further duplications and divergences have created a diverse range of α- and β-like globin genes that are regulated so that certain forms occur at different stages of development.

Iron's Oxidation State in Oxyhemoglobin

Assigning oxygenated hemoglobin's oxidation state is difficult because oxyhemoglobin (Hb-O_2), by experimental measurement, is diamagnetic (no net unpaired electrons), yet the lowest-energy (ground-state) electron configurations in both oxygen and iron are paramagnetic (suggesting at least one unpaired electron in the complex). The low-est-energy form of oxygen, and the lowest energy forms of the relevant oxidation states of iron, are these:

- Triplet oxygen, the lowest-energy molecular oxygen species, has two unpaired electrons in antibonding π^* molecular orbitals.

- Iron(II) tends to exist in a high-spin $3d^6$ configuration with four unpaired electrons.

- Iron(III) ($3d^5$) has an odd number of electrons, and thus must have one or more unpaired electrons, in any energy state.

All of these structures are paramagnetic (have unpaired electrons), not diamagnetic. Thus, a non-intuitive (e.g., a higher-energy for at least one species) distribution of electrons in the combination of iron and oxygen must exist, in order to explain the observed diamagnetism and no unpaired electrons.

The two logical possibilities to produce diamagnetic (no net spin) Hb-O_2 are:

1. Low-spin Fe^{2+} binds to singlet oxygen. Both low-spin iron and singlet oxygen are diamagnetic. However, the singlet form of oxygen is the higher-energy form of the molecule.

2. Low-spin Fe^{3+} binds to $O_2^{\cdot-}$ (the superoxide ion) and the two unpaired electrons couple antiferromagnetically, giving observed diamagnetic properties. Here, the iron has been oxidized (has lost one electron), and the oxygen has been reduced (has gained one electron).

Another possible model in which low-spin Fe^{4+} binds to peroxide, O_2^{2-}, can be ruled out by itself, because the iron is paramagnetic (although the peroxide ion is diamagnetic). Here, the iron has been oxidized by two electrons, and the oxygen reduced by two electrons.

Direct Experimental Data:

- X-ray photoelectron spectroscopy suggests iron has an oxidation state of approximately 3.2

- Infrared vibrational frequencies of the O-O bond suggests a bond length fitting with superoxide (a bond order of about 1.6, with superoxide being 1.5).

- X-ray Absorption Near Edge Structures at the iron K-edge. The energy shift of 5 eV between deoxyhemoglobin and oxyhemoglobin, as for all the methemoglobin species, strongly suggests an actual local charge closer to Fe^{3+} than Fe^{2+}.

Thus, the nearest formal oxidation state of iron in Hb-O_2 is the +3 state, with oxygen in the –1 state (as superoxide .O_2^-). The diamagnetism in this configuration arises from the single unpaired electron on superoxide aligning antiferromagnetically with the single unpaired electron on iron (in a low-spin d^5 state), to give no net spin to the entire configuration, in accordance with diamagnetic oxyhemoglobin from experiment.

The second choice of the logical possibilities above for diamagnetic oxyhemoglobin being found correct by experiment, is not surprising: singlet oxygen (possibility #1) is an unrealistically high energy state. Model 3 leads to unfavorable separation of charge (and does not agree with the magnetic data), although it could make a minor contribution as a resonance form. Iron's shift to a higher oxidation state in Hb-O_2 decreases the atom's size, and allows it into the plane of the porphyrin ring, pulling on the coordinated histidine residue and initiating the allosteric changes seen in the globulins.

Early postulates by bio-inorganic chemists claimed that possibility #1 (above) was correct and that iron should exist in oxidation state II. This conclusion seemed likely, since the iron oxidation state III as methemoglobin, when *not* accompanied by superoxide .O_2^- to "hold" the oxidation electron, was known to render hemoglobin incapable of binding normal triplet O_2 as it occurs in the air. It was thus assumed that iron remained as Fe(II) when oxygen gas was bound in the lungs. The iron chemistry in this previous classical model was elegant, but the required presence of the diamagnetic, high-energy, singlet oxygen molecule was never explained. It was classically argued that the binding of an oxygen molecule placed high-spin iron(II) in an octahedral field of strong-field ligands; this change in field would increase the crystal field splitting energy, causing iron's electrons to pair into the low-spin configuration, which would be diamagnetic in Fe(II). This forced low-spin pairing is indeed thought to happen in iron when oxygen binds, but is not enough to explain iron's change in size. Extraction of an additional electron from iron by oxygen is required to explain both iron's smaller size and observed increased oxidation state, and oxygen's weaker bond.

The assignment of a whole-number oxidation state is a formalism, as the covalent bonds are not required to have perfect bond orders involving whole electron transfer. Thus, all three models for paramagnetic Hb-O_2 may contribute to some small degree (by resonance) to the actual electronic configuration of Hb-O_2. However, the model of iron in Hb-O_2 being Fe(III) is more correct than the classical idea that it remains Fe(II).

Cooperativity

A schematic visual model of oxygen-binding process, showing all four monomers and hemes, and protein chains only as diagramatic coils, to facilitate visualization into the molecule. Oxygen is not shown in this model, but, for each of the iron atoms, it binds to the iron (red sphere) in the flat heme. For example, in the upper-left of the four hemes shown, oxygen binds at the left of the iron atom shown in the upper-left of diagram. This causes the iron atom to move backward into the heme that holds it (the iron moves upward as it binds oxygen, in this illustration), tugging the histidine residue (modeled as a red pentagon on the right of the iron) closer, as it does. This, in turn, pulls on the protein chain holding the histidine.

When oxygen binds to the iron complex, it causes the iron atom to move back toward the center of the plane of the porphyrin ring. At the same time, the imidazole side-chain of the histidine residue interacting at the other pole of the iron is pulled toward the porphyrin ring. This interaction forces the plane of the ring sideways toward the outside of the tetramer, and also induces a strain in the protein helix containing the histidine as it moves nearer to the iron atom. This strain is transmitted to the remaining three monomers in the tetramer, where it induces a similar conformational change in the other heme sites such that binding of oxygen to these sites becomes easier.

In the tetrameric form of normal adult hemoglobin, the binding of oxygen is, thus, a cooperative process. The binding affinity of hemoglobin for oxygen is increased by the oxygen saturation of the molecule, with the first oxygens bound influencing the shape of the binding sites for the next oxygens, in a way favorable for binding. This positive cooperative binding is achieved through steric conformational changes of the hemoglobin protein complex as discussed above; i.e., when one subunit protein in hemoglobin becomes oxygenated, a conformational or structural change in the whole complex is initiated, causing the other subunits to gain an increased affinity for oxygen. As a consequence, the oxygen binding curve of hemoglobin is sigmoidal, or S-shaped, as opposed to the normal hyperbolic curve associated with noncooperative binding.

The dynamic mechanism of the cooperativity in hemoglobin and its relation with the low-frequency resonance has been discussed.

Binding for Ligands other than Oxygen

Besides the oxygen ligand, which binds to hemoglobin in a cooperative manner, hemoglobin ligands also include competitive inhibitors such as carbon monoxide (CO) and allosteric ligands such as carbon dioxide (CO_2) and nitric oxide (NO). The carbon dioxide is bound to amino groups of the globin proteins as carbaminohemoglobin, and is thought to account for about 10% of carbon dioxide transport in mammals. Nitric oxide is bound to specific thiol groups in the globin protein to form an S-nitrosothiol, which dissociates into free nitric oxide and thiol again, as the hemoglobin releases oxygen from its heme site. This nitric oxide transport to peripheral tissues is hypothesized to assist oxygen transport in tissues, by releasing vasodilatory nitric oxide to tissues in which oxygen levels are low.

Competitive

The binding of oxygen is affected by molecules such as carbon monoxide (CO) (for example, from tobacco smoking, car exhaust, and incomplete combustion in furnaces).

CO competes with oxygen at the heme binding site. Hemoglobin binding affinity for CO is 250 times greater than its affinity for oxygen, meaning that small amounts of CO dramatically reduce hemoglobin's ability to transport oxygen. Since carbon monoxide is a colorless, odorless and tasteless gas, and poses a potentially fatal threat, detectors have become commercially available to warn of dangerous levels in residences. When hemoglobin combines with CO, it forms a very bright red compound called carboxyhemoglobin, which may cause the skin of CO poisoning victims to appear pink in death, instead of white or blue. When inspired air contains CO levels as low as 0.02%, headache and nausea occur; if the CO concentration is increased to 0.1%, unconsciousness will follow. In heavy smokers, up to 20% of the oxygen-active sites can be blocked by CO.

In similar fashion, hemoglobin also has competitive binding affinity for cyanide (CN^-), sulfur monoxide (SO), and sulfide (S^{2-}), including hydrogen sulfide (H_2S). All of these bind to iron in heme without changing its oxidation state, but they nevertheless inhibit oxygen-binding, causing grave toxicity.

The iron atom in the heme group must initially be in the ferrous (Fe^{2+}) oxidation state to support oxygen and other gases' binding and transport (it temporarily switches to ferric during the time oxygen is bound, as explained above). Initial oxidation to the ferric (Fe^{3+}) state without oxygen converts hemoglobin into "hemiglobin" or methemoglobin (pronounced "MET-hemoglobin"), which cannot bind oxygen. Hemoglobin in normal red blood cells is protected by a reduction system to keep this from happening. Nitric oxide is capable of converting a small fraction of hemoglobin to methemoglobin in red blood cells. The latter reaction is a remnant activity of the more ancient nitric oxide dioxygenase function of globins.

Allosteric

Carbon *dioxide* occupies a different binding site on the hemoglobin. Carbon dioxide is more readily dissolved in deoxygenated blood, facilitating its removal from the body after the oxygen has been released to tissues undergoing metabolism. This increased affinity for carbon dioxide by the venous blood is known as the Haldane effect. Through the enzyme carbonic anhydrase, carbon dioxide reacts with water to give carbonic acid, which decomposes into bicarbonate and protons:

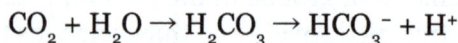

$$CO_2 + H_2O \rightarrow H_2CO_3 \rightarrow HCO_3^- + H^+$$

The sigmoidal shape of hemoglobin's oxygen-dissociation curve results from cooperative binding of oxygen to hemoglobin.

Hence, blood with high carbon dioxide levels is also lower in pH (more acidic). Hemoglobin can bind protons and carbon dioxide, which causes a conformational change in the protein and facilitates the release of oxygen. Protons bind at various places on the protein, while carbon dioxide binds at the α-amino group. Carbon dioxide binds to hemoglobin and forms carbaminohemoglobin. This decrease in hemoglobin's affinity for oxygen by the binding of carbon dioxide and acid is known as the Bohr effect (shifts the O_2-saturation curve to the *right*). Conversely, when the carbon dioxide levels in the blood decrease (i.e., in the lung capillaries), carbon dioxide and protons are released from hemoglobin, increasing the oxygen affinity of the protein. A reduction in the total binding capacity of hemoglobin to oxygen (i.e. shifting the curve down, not just to the right) due to reduced pH is called the root effect. This is seen in bony fish.

It is necessary for hemoglobin to release the oxygen that it binds; if not, there is no point in binding it. The sigmoidal curve of hemoglobin makes it efficient in binding (taking up O_2 in lungs), and efficient in unloading (unloading O_2 in tissues).

In people acclimated to high altitudes, the concentration of 2,3-Bisphosphoglycerate (2,3-BPG) in the blood is increased, which allows these individuals to deliver a larger amount of oxygen to tissues under conditions of lower oxygen tension. This phenomenon, where molecule Y affects the binding of molecule X to a transport molecule Z, is called a *heterotropic*allosteric effect. Hemoglobin in organisms at high altitudes has also adapted such that it has less of an affinity for 2,3-BPG and so the protein will be shifted more towards its R state. In its R state, hemoglobin will bind oxygen more readily, thus allowing organisms to perform the necessary metabolic processes when oxygen is present at low partial pressures.

Animals other than humans use different molecules to bind to hemoglobin and change its O_2 affinity under unfavorable conditions. Fish use both ATP and GTP. These bind to a phosphate "pocket" on the fish hemoglobin molecule, which stabilizes the tense state and therefore decreases oxygen affinity. GTP reduces hemoglobin oxygen affinity much more than ATP, which is thought to be due to an extra hydrogen bond formed that further stabilizes the tense state. Under hypoxic conditions, the concentration of both ATP and GTP is reduced in fish red blood cells to increase oxygen affinity.

A variant hemoglobin, called fetal hemoglobin (HbF, $\alpha_2\gamma_2$), is found in the developing fetus, and binds oxygen with greater affinity than adult hemoglobin. This means that the oxygen binding curve for fetal hemoglobin is left-shifted (i.e., a higher percentage of hemoglobin has oxygen bound to it at lower oxygen tension), in comparison to that of adult hemoglobin. As a result, fetal blood in the placenta is able to take oxygen from maternal blood.

Hemoglobin also carries nitric oxide (NO) in the globin part of the molecule. This improves oxygen delivery in the periphery and contributes to the control of respiration. NO binds reversibly to a specific cysteine residue in globin; the binding depends on the state (R or T) of the hemoglobin. The resulting S-nitrosylated hemoglobin influences various NO-related activities such as the control of vascular resistance, blood pressure and respiration. NO is not released in the cytoplasm of erythrocytes but transported by an anion exchanger called AE1 out of them.

Types in Humans

Hemoglobin variants are a part of the normal embryonic and fetal development. They may also be pathologic mutant forms of hemoglobin in a population, caused by variations in genetics. Some well-known hemoglobin variants, such as sickle-cell anemia, are responsible for diseases and are considered hemoglobinopathies. Other variants cause no detectable pathology, and are thus considered non-pathological variants.

In the embryo:

- Gower 1 ($\zeta_2\varepsilon_2$)

- Gower 2 ($\alpha_2\varepsilon_2$) (PDB: 1A9W)

- Hemoglobin Portland I ($\zeta_2\gamma_2$)

- Hemoglobin Portland II ($\zeta_2\beta_2$).

In the fetus:

- Hemoglobin F ($\alpha_2\gamma_2$) (PDB: 1FDH).

After birth:

- Hemoglobin A ($\alpha_2\beta_2$) (PDB: 1BZ0) – The most common with a normal amount over 95%

- Hemoglobin A_2 ($\alpha_2\delta_2$) – δ chain synthesis begins late in the third trimester and, in adults, it has a normal range of 1.5–3.5%

- Hemoglobin F ($\alpha_2\gamma_2$) – In adults Hemoglobin F is restricted to a limited population of red cells called F-cells. However, the level of Hb F can be elevated in persons with sickle-cell disease and beta-thalassemia.

Gene expression of hemoglobin before and after birth. Also identifies the types of cells and organs in which the gene expression (data on *Wood W.G.*, (1976). Br. Med. Bull. 32, 282.)

Variant forms that cause disease:

- Hemoglobin D-Punjab – ($\alpha_2\beta^D_2$) – A variant form of hemoglobin.

- Hemoglobin H (β_4) – A variant form of hemoglobin, formed by a tetramer of β chains, which may be present in variants of α thalassemia.

- Hemoglobin Barts (γ_4) – A variant form of hemoglobin, formed by a tetramer of γ chains, which may be present in variants of α thalassemia.

- Hemoglobin S ($\alpha_2\beta^S_2$) – A variant form of hemoglobin found in people with sickle cell disease. There is a variation in the β-chain gene, causing a change in the properties of hemoglobin, which results in sickling of red blood cells.

- Hemoglobin C ($\alpha_2\beta^C_2$) – Another variant due to a variation in the β-chain gene. This variant causes a mild chronic hemolytic anemia.

- Hemoglobin E ($\alpha_2\beta^E_2$) – Another variant due to a variation in the β-chain gene. This variant causes a mild chronic hemolytic anemia.

- Hemoglobin AS – A heterozygous form causing sickle cell trait with one adult gene and one sickle cell disease gene

- Hemoglobin SC disease – A compound heterozygous form with one sickle gene and another encoding Hemoglobin C.

Degradation in Vertebrate Animals

When red cells reach the end of their life due to aging or defects, they are removed from the circulation by the phagocytic activity of macrophages in the spleen or the liver or hemolyze within the circulation. Free hemoglobin is then cleared from the circulation via the hemoglobin transporter CD163, which is exclusively expressed on monocytes or macrophages. Within these cells the hemoglobin molecule is broken up, and the iron gets recycled. This process also produces one molecule of carbon monoxide for every molecule of heme degraded. Heme degradation is one of the few natural sources of carbon monoxide in the human body, and is responsible for the normal blood levels of carbon monoxide even in people breathing pure air. The other major final product of heme degradation is bilirubin. Increased levels of this chemical are detected in the blood if red cells are being destroyed more rapidly than usual. Improperly degraded hemoglobin protein or hemoglobin that has been released from the blood cells too rapidly can clog small blood vessels, especially the delicate blood filtering vessels of the kidneys, causing kidney damage. Iron is removed from heme and salvaged for later use, it is stored as hemosiderin or ferritin in tissues and transported in plasma by beta globulins as transferrins. When the porphyrin ring is broken up, the fragments are normally secreted as a yellow pigment called bilirubin, which is secreted into the intestines as bile. Intestines metabolise bilirubin into urobilinogen. Urobilinogen leaves the body in faeces, in a pigment called stercobilin. Globulin is metabolised into amino acids that are then released into circulation.

Role in Disease

Hemoglobin deficiency can be caused either by a decreased amount of hemoglobin molecules, as in anemia, or by decreased ability of each molecule to bind oxygen at the same partial pressure of oxygen. Hemoglobinopathies (genetic defects resulting in abnormal structure of the hemoglobin molecule) may cause both. In any case, hemoglobin deficiency decreases blood oxygen-carrying capacity. Hemoglobin deficiency is, in general, strictly distinguished from hypoxemia, defined as decreased partial pressure of oxygen in blood, although both are causes of hypoxia (insufficient oxygen supply to tissues).

Other common causes of low hemoglobin include loss of blood, nutritional deficiency, bone marrow problems, chemotherapy, kidney failure, or abnormal hemoglobin (such as that of sickle-cell disease).

The ability of each hemoglobin molecule to carry oxygen is normally modified by altered blood pH or CO_2, causing an altered oxygen–hemoglobin dissociation curve. However, it can also be pathologically altered in, e.g., carbon monoxide poisoning.

Decrease of hemoglobin, with or without an absolute decrease of red blood cells, leads to symptoms of anemia. Anemia has many different causes, although iron deficiency and its resultant iron deficiency anemia are the most common causes in the Western world. As absence of iron decreases heme synthesis, red blood cells in iron deficiency anemia are *hypochromic* (lacking the red hemoglobin pigment) and *microcytic* (smaller than normal). Other anemias are rarer. In hemolysis (accelerated breakdown of red blood cells), associated jaundice is caused by the hemoglobin metabolite bilirubin, and the circulating hemoglobin can cause renal failure.

Some mutations in the globin chain are associated with the hemoglobinopathies, such as sickle-cell disease and thalassemia. Other mutations, as discussed at the beginning of the article, are benign and are referred to merely as hemoglobin variants.

There is a group of genetic disorders, known as the *porphyrias* that are characterized by errors in metabolic pathways of heme synthesis. King George III of the United Kingdom was probably the most famous porphyria sufferer.

To a small extent, hemoglobin A slowly combines with glucose at the terminal valine (an alpha aminoacid) of each β chain. The resulting molecule is often referred to as Hb A_{1c}. As the concentration of glucose in the blood increases, the percentage of Hb A that turns into Hb A_{1c} increases. In diabetics whose glucose usually runs high, the percent Hb A_{1c} also runs high. Because of the slow rate of Hb A combination with glucose, the Hb A_{1c} percentage is representative of glucose level in the blood averaged over a longer time (the half-life of red blood cells, which is typically 50–55 days).

Glycosylated hemoglobin is the form of hemoglobin to which glucose is bound. The binding of glucose to amino acids in the hemoglobin takes place spontaneously (without the help of an enzyme) in many proteins, and is not known to serve a useful purpose. However, the binding to hemoglobin does serve as a record for average blood glucose levels over the lifetime of red cells, which is approximately 120 days. The levels of glycosylated hemoglobin are therefore measured in order to monitor the long-term control of the chronic disease of type 2 diabetes mellitus (T2DM). Poor control of T2DM results in high levels of glycosylated hemoglobin in the red blood cells. The normal reference range is approximately 4–5.9 %. Though difficult to obtain, values less than 7% are recommended for people with T2DM. Levels greater than 9% are associated with poor control of the glycosylated hemoglobin, and levels greater than 12% are associated with very poor control. Diabetics who keep their glycosylated hemoglobin levels close to 7% have a much better chance of avoiding the complications that may accompany diabetes (than those whose levels are 8% or higher). In addition, increased glycosylation of hemoglobin increases its affinity for oxygen, therefore preventing its release at the tissue and inducing a level of hypoxia in extreme cases.

Elevated levels of hemoglobin are associated with increased numbers or sizes of red blood cells, called polycythemia. This elevation may be caused by congenital heart disease, cor pulmonale, pulmonary fibrosis, too much erythropoietin, or polycythemia vera. High hemoglobin levels may also be caused by exposure to high altitudes, smoking, dehydration (artificially by concentrating Hb), advanced lung disease and certain tumors.

A recent study done in Pondicherry, India, shows its importance in coronary artery disease.

Diagnostic Uses

Hemoglobin concentration measurement is among the most commonly performed blood tests, usually as part of a complete blood count. For example, it is typically tested before or after blood donation. Results are reported in g/L, g/dL or mol/L. 1 g/dL equals about 0.6206 mmol/L, although the latter units are not used as often due to uncertainty regarding the polymeric state of the molecule. This conversion factor, using the single globin unit molecular weight of 16,000 Da, is more common for hemoglobin concentration in blood. For MCHC (mean corpuscular hemoglobin concentration) the conversion factor 0.155, which uses the tetramer weight of 64,500 Da, is more common. Normal levels are:

- Men: 13.8 to 18.0 g/dL (138 to 180 g/L, or 8.56 to 11.17 mmol/L)

- Women: 12.1 to 15.1 g/dL (121 to 151 g/L, or 7.51 to 9.37 mmol/L)

- Children: 11 to 16 g/dL (111 to 160 g/L, or 6.83 to 9.93 mmol/L)

- Pregnant women: 11 to 14 g/dL (110 to 140 g/L, or 6.83 to 8.69 mmol/L) (9.5 to 15 usual value during pregnancy)

Normal values of hemoglobin in the 1st and 3rd trimesters of pregnant women must be at least 11 g/dL and at least 10.5 g/dL during the 2nd trimester.

Dehydration or hyperhydration can greatly influence measured hemoglobin levels. Albumin can indicate hydration status.

If the concentration is below normal, this is called anemia. Anemias are classified by the size of red blood cells, the cells that contain hemoglobin in vertebrates. The anemia is called "microcytic" if red cells are small, "macrocytic" if they are large, and "normocytic" otherwise.

Hematocrit, the proportion of blood volume occupied by red blood cells, is typically about three times the hemoglobin concentration measured in g/dL. For example, if the hemoglobin is measured at 17 g/dL, that compares with a hematocrit of 51%.

Laboratory hemoglobin test methods require a blood sample (arterial, venous, or capillary) and analysis on hematology analyzer and CO-oximeter. Additionally, a new noninvasive hemoglobin (SpHb) test method called Pulse CO-Oximetry is also available with comparable accuracy to invasive methods.

Concentrations of oxy- and deoxyhemoglobin can be measured continuously, regionally and noninvasively using NIRS. NIRS can be used both on the head as on muscles. This technique is often used for research in e.g. elite sports training, ergonomics, rehabilitation, patient monitoring, neonatal research, functional brain monitoring, brain computer interface, urology (bladder contraction), neurology (Neurovascular coupling) and more.

Long-term control of blood sugar concentration can be measured by the concentration of Hb A_{1c}. Measuring it directly would require many samples because blood sugar levels vary widely through the day. Hb A_{1c} is the product of the irreversible reaction of hemoglobin A with glucose. A higher glucose concentration results in more Hb A_{1c}. Because the reaction is slow, the Hb A_{1c} proportion represents glucose level in blood averaged over the half-life of red blood cells, is typically 50–55 days. An Hb A_{1c} proportion of 6.0% or less show good long-term glucose control, while values above 7.0% are elevated. This test is especially useful for diabetics.

The functional magnetic resonance imaging (fMRI) machine uses the signal from deoxyhemoglobin, which is sensitive to magnetic fields since it is paramagnetic. Combined measurement with NIRS shows good correlation with both the oxy- and deoxyhemoglobin signal compared to the BOLD signal.

Athletic Tracking and Self Tracking Uses

Hemoglobin can be tracked noninvasively, to build an individual data set tracking the hemoconcentration and hemodilution effects of daily activities for better understanding of sports performance and training. Athletes are often concerned about endurance and intensity of exercise. Using the scientific technique of absorption spectroscopy with eight wavelengths of light. This method is similar to a pulse oximeter, which consists of a small sensing device that clips to the finger. The sensor uses light-emitting diodes that emit red and infrared light through the tissue to a light detector, which then sends a signal to a processor to calculate the absorption of light by the hemoglobin protein.

Analogues in Non-vertebrate Organisms

A variety of oxygen-transport and -binding proteins exist in organisms throughout the animal and plant kingdoms. Organisms including bacteria, protozoans, and fungi all have hemoglobin-like proteins whose known and predicted roles include the reversible binding of gaseous ligands. Since many of these proteins contain globins and the heme moiety (iron in a flat porphyrin support), they are often called hemoglobins, even if their overall tertiary structure is very different from that of vertebrate hemoglobin. In particular, the distinction of "myoglobin" and hemoglobin in lower animals is often impossible, because some of these organisms do not contain muscles. Or, they may have a recognizable separate circulatory system but not one that deals with oxygen transport (for example, many insects and other arthropods). In all these groups, heme/globin-containing molecules (even monomeric globin ones) that deal with gas-binding are referred to as oxyhemoglobins. In addition to dealing with transport and sensing of oxygen, they may also deal with NO, CO_2, sulfide compounds, and even O_2 scavenging in environments that must be anaerobic. They may even deal with detoxification of chlorinated materials in a way analogous to heme-containing P450 enzymes and peroxidases.

The giant tube worm *Riftia pachyptila* showing red hemoglobin-containing plumes

The structure of hemoglobins varies across species. Hemoglobin occurs in all kingdoms of organisms, but not in all organisms. Primitive species such as bacteria, protozoa, algae, and plants often have single-globin hemoglobins. Many nematode worms, molluscs, and crustaceans contain very large multisubunit molecules, much larger than those in vertebrates. In particular, chimeric hemoglobins found in fungi and giant annelids may contain both globin and other types of proteins.

One of the most striking occurrences and uses of hemoglobin in organisms is in the giant tube worm (*Riftia pachyptila*, also called Vestimentifera), which can reach 2.4 meters length and populates ocean volcanic vents. Instead of a digestive tract, these worms contain a population of bacteria constituting half the organism's weight. The bacteria react with H_2S from the vent and O_2 from the water to produce energy to make food from H_2O and CO_2. The worms end with a deep-red fan-like structure ("plume"), which extends into the water and absorbs H_2S and O_2 for the bacteria, and CO_2 for use as synthetic raw material similar to photosynthetic plants. The structures are bright-red due to their containing several extraordinarily complex hemoglobins that have up to 144 globin chains, each including associated heme structures. These hemoglobins are remarkable for being able to carry oxygen in the presence of sulfide, and even to carry sulfide, without being completely "poisoned" or inhibited by it as hemoglobins in most other species are.

Other Oxygen-binding Proteins

Myoglobin

> Found in the muscle tissue of many vertebrates, including humans, it gives muscle tissue a distinct red or dark gray color. It is very similar to hemoglobin in structure and sequence, but is not a tetramer; instead, it is a monomer that lacks cooperative binding. It is used to store oxygen rather than transport it.

Hemocyanin

The second most common oxygen-transporting protein found in nature, it is found in the blood of many arthropods and molluscs. Uses copper prosthetic groups instead of iron heme groups and is blue in color when oxygenated.

Hemerythrin

Some marine invertebrates and a few species of annelid use this iron-containing non-heme protein to carry oxygen in their blood. Appears pink/violet when oxygenated, clear when not.

Chlorocruorin

Found in many annelids, it is very similar to erythrocruorin, but the heme group is significantly different in structure. Appears green when deoxygenated and red when oxygenated.

Vanabins

Also known as *vanadium chromagens*, they are found in the blood of sea squirts. They were once hypothesized to use the rare metal vanadium as an oxygen binding prosthetic group. However, although they do contain vanadium by preference, they apparently bind little oxygen, and thus have some other function, which has not been elucidated (sea squirts also contain some hemoglobin). They may act as toxins.

Erythrocruorin

Found in many annelids, including earthworms, it is a giant free-floating blood protein containing many dozens—possibly hundreds—of iron- and heme-bearing protein subunits bound together into a single protein complex with a molecular mass greater than 3.5 million daltons.

Pinnaglobin

Only seen in the mollusc *Pinna nobilis*. Brown manganese-based porphyrin protein.

Leghemoglobin

In leguminous plants, such as alfalfa or soybeans, the nitrogen fixing bacteria in the roots are protected from oxygen by this iron heme containing oxygen-binding protein. The specific enzyme protected is nitrogenase, which is unable to reduce nitrogen gas in the presence of free oxygen.

Coboglobin

A synthetic cobalt-based porphyrin. Coboprotein would appear colorless when oxygenated, but yellow when in veins.

Presence in Nonerythroid Cells

Some nonerythroid cells (i.e., cells other than the red blood cell line) contain hemoglobin. In the brain, these include the A9 dopaminergic neurons in the substantia nigra, astrocytes in the cerebral cortex and hippocampus, and in all mature oligodendrocytes. It has been suggested that brain hemoglobin in these cells may enable the "storage of oxygen to provide a homeostatic mechanism in anoxic conditions, which is especially important for A9 DA neurons that have an elevated metabolism with a high requirement for energy production". It has been noted further that "A9 dopaminergic neurons may be at particular risk since in addition to their high mitochondrial activity they are under intense oxidative stress caused by the production of hydrogen peroxide via autoxidation and/or monoamine oxidase (MAO)-mediated deamination of dopamine and the subsequent reaction of accessible ferrous iron to generate highly toxic hydroxyl radicals". This may explain the risk of these cells for degeneration in Parkinson's disease. The hemoglobin-derived iron in these cells is not the cause of the post-mortem darkness of these cells (origin of the Latin name, substantia *nigra*), but rather is due to neuromelanin.

Outside the brain, hemoglobin has non-oxygen-carrying functions as an antioxidant and a regulator of iron metabolism in macrophages,alveolar cells, and mesangial cells in the kidney.

In History, Art and Music

Heart of Steel (Hemoglobin) (2005) by Julian Voss-Andreae. The images show the 5-foot (1.60 m) tall sculpture right after installation, after 10 days, and after several months of exposure to the elements.

Historically, an association between the color of blood and rust occurs in the association of the planet Mars, with the Roman god of war, since the planet is an orange-red, which reminded the ancients of blood. Although the color of the planet is due to iron compounds in combination with oxygen in the Martian soil, it is a common misconception that the iron in hemoglobin and its oxides gives blood its red color. The color is actually due to the porphyrinmoiety of hemoglobin to which the iron is bound, not the iron itself, although the ligation and redox state of the iron can influence the pi to pi* or n to pi* electronic transitions of the porphyrin and hence its optical characteristics.

Artist Julian Voss-Andreae created a sculpture called "Heart of Steel (Hemoglobin)" in 2005, based on the protein's backbone. The sculpture was made from glass and weathering steel. The intentional rusting of the initially shiny work of art mirrors hemoglobin's fundamental chemical reaction of oxygen binding to iron.

Montreal artist Nicolas Baier created Lustre (Hémoglobine), a sculpture in stainless steel that shows the structure of the hemoglobin molecule. It is displayed in the atrium of McGill University Health Centre's research centre in Montreal. The sculpture measures about 10 metres × 10 metres × 10 metres.

Pancreatic Ribonuclease

Pancreatic ribonucleases (EC3.1.27.5, *RNase, RNase I, RNase A, pancreatic RNase, ribonuclease I, endoribonuclease I, ribonucleic phosphatase, alkaline ribonuclease, ribonuclease, gene S glycoproteins, Ceratitis capitata alkaline ribonuclease, SLSG glycoproteins, gene S locus-specific glycoproteins, S-genotype-assocd. glycoproteins, ribonucleate 3'-pyrimidino-oligonucleotidohydrolase*) are pyrimidine-specific endonucleases found in high quantity in the pancreas of certain mammals and of some reptiles.

Specifically, the enzymes are involved in endonucleolytic cleavage of 3'-phosphomononucleotides and 3'-phosphooligonucleotides ending in C-P or U-P with 2',3'-cyclic phosphate intermediates. Ribonuclease can unwind the RNA helix by complexing with single-stranded RNA; the complex arises by an extended multi-site cation-anion interaction between lysine and arginine residues of the enzyme and phosphate groups of the nucleotides.

Notable Family Members

Bovine pancreatic ribonuclease is the best-studied member of the family and has served as a model system in work related to protein folding, disulfide bond formation, protein crystallography and spectroscopy, and protein dynamics.

Other proteins belonging to the pancreatic ribonuclease superfamily include: bovine seminal vesicle and brain ribonucleases; kidney non-secretory ribonucleases; liver-type ribonucleases;angiogenin, which induces vascularisation of normal and malignant tissues; eosinophil cationic protein, a cytotoxin and helminthotoxin with ribonuclease activity; and frog liver ribonuclease and frog sialic acid-binding lectin. The sequence of pancreatic ribonucleases contains four conserved disulfide bonds and three amino acid residues involved in the catalytic activity.

Human Genes

Humangenes encoding proteins containing this domain include:

- ANG,
- RNASE1, RNASE10, RNASE12, RNASE2, RNASE3, RNASE4, RNASE6, RNASE7, and RNASE8.

Cytotoxicity

Some members of the pancreatic ribonuclease family have cytotoxic effects. Mammalian cells are protected from these effects due to their extremely high affinity for ribonuclease inhibitor (RI), which protects cellular RNA from degradation by pancreatic ribonucleases. Pancreatic ribonucleases that are not inhibited by RI are approximately as toxic as alpha-sarcin, diphtheria toxin, or ricin.

Two pancreatic ribonucleases isolated from the oocytes of the Northern leopard frog - amphinase and ranpirnase - are not inhibited by RI and show differential cytotoxicity against tumor cells. Ranpirnase was studied in a Phase IIIclinical trial as a treatment candidate for mesothelioma, but the trial did not demonstrate statistical significance against primary endpoints.

Antibody

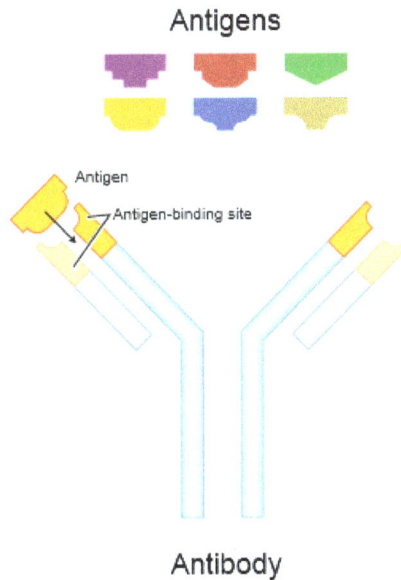

Each antibody binds to a specific antigen; an interaction similar to a lock and key.

An antibody (Ab), also known as an immunoglobulin (Ig), is a large, Y-shaped protein produced mainly by plasma cells that is used by the immune system to identify and neutralize pathogens such as bacteria and viruses. The antibody recognizes a unique molecule of the harmful agent, called an antigen, via the Fab's variable region. Each tip of the "Y" of an antibody contains a paratope (analogous to a lock) that is specific for one particular epitope (similarly analogous to a key) on an antigen, allowing these two structures to bind together with precision. Using this binding mechanism, an antibody can *tag* a microbe or an infected cell for attack by other parts of the immune system, or can neutralize its target directly (for example, by blocking a part of a microbe that is essential for its invasion and survival). Depending on the antigen, the binding may impede the biological process causing the disease or may activate macrophages to destroy the foreign substance. The ability of an antibody to communicate with the other components of the immune system is mediated via its Fc region (located at the base of the "Y"), which contains a conserved glycosylation site involved in these interactions. The production of antibodies is the main function of the humoral immune system.

Antibodies are secreted by B cells of the adaptive immune system, mostly by differentiated B cells called plasma cells. Antibodies can occur in two physical forms, a soluble form that is secreted from the cell to be free in the blood plasma, and a membrane-bound form that is attached to the surface of a B cell and is referred to as the B-cell receptor (BCR). The BCR is found only on the surface of B cells and facilitates the activation of these cells and their subsequent differentiation into either antibody factories called plasma cells or memory B cells that will survive in the body and remember that same antigen so the B cells can respond faster upon future exposure. In most cases, interaction of the B cell with a T helper cell is necessary to produce full activation of the B cell and, therefore, antibody generation following antigen binding. Soluble antibodies are released into the blood and tissue fluids, as well as many secretions to continue to survey for invading microorganisms.

Antibodies are glycoproteins belonging to the immunoglobulin superfamily. They constitute most of the gamma globulin fraction of the blood proteins. They are typically made of basic structural units—each with two large heavy chains and two small light chains. There are several different types of antibody heavy chains that define the five different types of crystallisable fragments (Fc) that may be attached to the antigen-binding fragments. The five different types of Fc regions allow antibodies to be grouped into five *isotypes*. Each Fc region of a particular antibody isotype is able to bind to its specific Fc Receptor (except for IgD, which is essentially the BCR), thus allowing the antigen-antibody complex to mediate different roles depending on which FcR it binds. The ability of an antibody to bind to its corresponding FcR is further modulated by the structure of the glycan(s) present at conserved sites within its Fc region. The ability of antibodies to bind to FcRs helps to direct the appropriate immune response for each different type of foreign object they encounter. For example, IgE is responsible for an allergic response consisting of mast cell degranulation and histamine release. IgE's Fab paratope binds to allergic antigen, for example house dust mite particles, while its Fc region binds to Fc receptor ε. The allergen-IgE-FcRε interaction mediates allergic signal transduction to induce conditions such as asthma.

Though the general structure of all antibodies is very similar, a small region at the tip of the protein is extremely variable, allowing millions of antibodies with slightly different tip structures, or antigen-binding sites, to exist. This region is known as the *hypervariable region*. Each of these variants can bind to a different antigen. This enormous diversity of antibody paratopes on the antigen-binding fragments allows the immune system to recognize an equally wide variety of antigens. The large and diverse population of antibody paratope is generated by random recombination events of a set of gene segments that encode different antigen-binding sites (or *paratopes*), followed by random mutations in this area of the antibody gene, which create further diversity. This recombinational process that produces clonal antibody paratope diversity is called V(D)

J or VJ recombination. Basically, the antibody paratope is polygenic, made up of three genes, V, D, and J. Each paratope locus is also polymorphic, such that during antibody production, one allele of V, one of D, and one of J is chosen. These gene segments are then joined together using random genetic recombination to produce the paratope. The regions where the genes are randomly recombined together is the hyper variable region used to recognise different antigens on a clonal basis.

Antibody genes also re-organize in a process called class switching that changes the one type of heavy chain Fc fragment to another, creating a different isotype of the antibody that retains the antigen-specific variable region. This allows a single antibody to be used by different types of Fc receptors, expressed on different parts of the immune system.

Forms

The membrane-bound form of an antibody may be called a *surface immunoglobulin* (sIg) or a *membrane immunoglobulin* (mIg). It is part of the *B cell receptor* (BCR), which allows a B cell to detect when a specific antigen is present in the body and triggers B cell activation. The BCR is composed of surface-bound IgD or IgM antibodies and associated Ig-α and Ig-β heterodimers, which are capable of signal transduction. A typical human B cell will have 50,000 to 100,000 antibodies bound to its surface. Upon antigen binding, they cluster in large patches, which can exceed 1 micrometer in diameter, on lipid rafts that isolate the BCRs from most other cell signaling receptors. These patches may improve the efficiency of the cellular immune response. In humans, the cell surface is bare around the B cell receptors for several hundred nanometers, which further isolates the BCRs from competing influences.

Antibody–antigen Interactions

The antibody's paratope interacts with the antigen's epitope. An antigen usually contains different epitopes along its surface arranged discontinuously, and dominant epitopes on a given antigen are called determinants.

Antibody and antigen interact by spatial complementarity (lock and key). The molecular forces involved in the Fab-epitope interaction are weak and non-specific – for example electrostatic forces, hydrogen bonds, hydrophobic interactions, and van der Waals forces. This means binding between antibody and antigen is reversible, and the antibody's affinity towards an antigen is relative rather than absolute. Relatively weak binding also means it is possible for an antibody to cross-react with different antigens of different relative affinities.

Often, once an antibody and antigen bind, they become an immune complex, which functions as a unitary object and can act as an antigen in its own right, being countered by other antibodies. Similarly, haptens are small molecules that provoke no immune response by themselves, but once they bind to proteins, the resulting complex or hapten-carrier adduct is antigenic.

Isotypes

Antibodies can come in different varieties known as isotypes or classes. In placental mammals there are five antibody isotypes known as IgA, IgD, IgE, IgG, and IgM. They are each named with an "Ig" prefix that stands for immunoglobulin, a name sometimes used interchangeably with antibody, and differ in their biological properties, functional locations and ability to deal with different antigens, as depicted in the table. The different suffixes of the antibody isotypes denote the different types of heavy chains the antibody contains, with each heavy chain class named alphabetically: α (alpha), γ (gamma), δ (delta), ε (epsilon), and μ (mu). This gives rise to IgA, IgG, IgD, IgE, and IgM, respectively.

Antibody isotypes of mammals			
Name	**Types**	**Description**	**Antibody complexes**
IgA	2	Found in mucosal areas, such as the gut, respiratory tract and urogenital tract, and prevents colonization by pathogens. Also found in saliva, tears, and breast milk.	
IgD	1	Functions mainly as an antigen receptor on B cells that have not been exposed to antigens. It has been shown to activate basophils and mast cells to produce antimicrobial factors.	Monomer IgD, IgE, IgG
IgE	1	Binds to allergens and triggers histamine release from mast cells and basophils, and is involved in allergy. Also protects against parasitic worms.	Dimer IgA
IgG	4	In its four forms, provides the majority of antibody-based immunity against invading pathogens. The only antibody capable of crossing the placenta to give passive immunity to the fetus.	Pentamer IgM
IgM	1	Expressed on the surface of B cells (monomer) and in a secreted form (pentamer) with very high avidity. Eliminates pathogens in the early stages of B cell-mediated (humoral) immunity before there is sufficient IgG.	

The antibody isotype of a B cell changes during cell development and activation. Immature B cells, which have never been exposed to an antigen, express only the IgM isotype in a cell surface bound form. The B lymphocyte, in this ready-to-respond form, is known as a "naive B lymphocyte." The naive B lymphocyte expresses both surface IgM and IgD. The co-expression of both of these immunoglobulin isotypes renders the B cell ready to respond to antigen. B cell activation follows engagement of the cell-bound antibody molecule with an antigen, causing the cell to divide and differentiate into an antibody-producing cell called a plasma cell. In this activated form, the B cell starts to produce antibody in a secreted form rather than a membrane-bound form. Some daughter cells of the activated B cells undergo isotype switching, a mechanism that causes the production of antibodies to change from IgM or IgD to the other antibody isotypes, IgE, IgA, or IgG, that have defined roles in the immune system.

Antibody isotypes not found in mammals		
Name	**Types**	**Description**
IgY		Found in birds and reptiles; related to mammalian IgG.
IgW		Found in sharks and skates; related to mammalian IgD.

Structure

Antibodies are heavy (~150 kDa) globularplasma proteins. They have sugar chains (glycans) added to conserved amino acid residues. In other words, antibodies are *glycoproteins*. The attached glycans are critically important to the structure and function of the antibody. Among other things the expressed glycans can modulate an antibody's affinity for its corresponding FcR(s).

The basic functional unit of each antibody is an immunoglobulin (Ig) monomer (containing only one Ig unit); secreted antibodies can also be dimeric with two Ig units as with IgA, tetrameric with four Ig units like teleost fish IgM, or pentameric with five Ig units, like mammalian IgM.

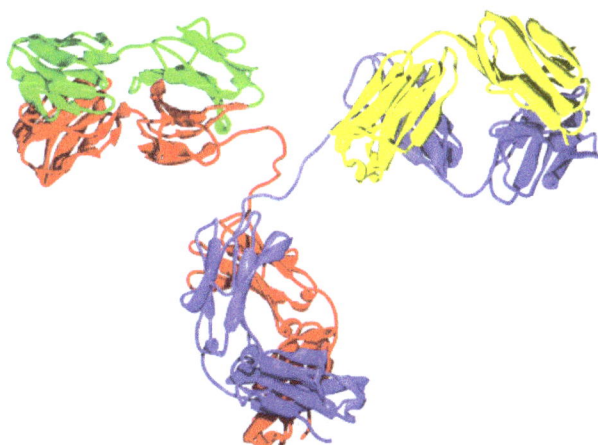

Several immunoglobulin domains make up the two heavy chains (red and blue) and the two light chains (green and yellow) of an antibody. The immunoglobulin domains are composed of between 7 (for constant domains) and 9 (for variable domains) β-strands.

The variable parts of an antibody are its V regions, and the constant part is its C region.

Immunoglobulin Domains

The Ig monomer is a "Y"-shaped molecule that consists of four polypeptide chains; two identical *heavy chains* and two identical *light chains* connected by disulfide bonds. Each chain is composed of structural domains called immunoglobulin domains. These domains contain about 70–110 amino acids and are classified into different categories (for example, variable or IgV, and constant or IgC) according to their size and function. They have a characteristic immunoglobulin fold in which two beta sheets create a "sandwich" shape, held together by interactions between conserved cysteines and other charged amino acids.

Heavy Chain

There are five types of mammalian Ig heavy chain denoted by the Greek letters: α, δ, ε, γ, and μ. The type of heavy chain present defines the *class* of antibody; these chains are found in IgA, IgD, IgE, IgG, and IgM antibodies, respectively. Distinct heavy chains differ in size and composition; α and γ contain approximately 450 amino acids, whereas μ and ε have approximately 550 amino acids.

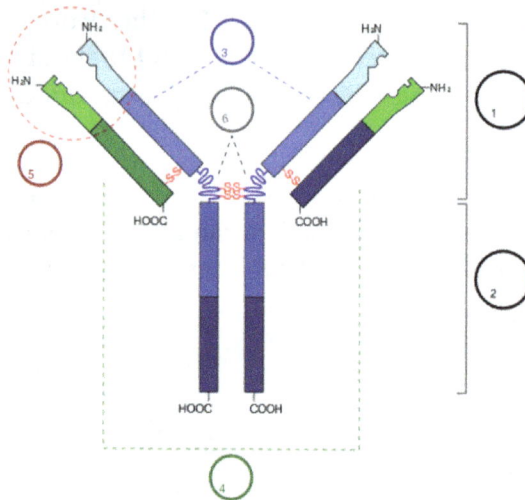

1. Fab region

2. Fc region

3. Heavy chain (blue) with one variable (V_H) domain followed by a constant domain (C_H1), a hinge region, and two more constant (C_H2 and C_H3) domains

4. Light chain (green) with one variable (V_L) and one constant (C_L) domain

5. Antigen binding site (paratope)

6. Hinge regions

Each heavy chain has two regions, the *constant region* and the *variable region*. The constant region is identical in all antibodies of the same isotype, but differs in antibodies of different isotypes. Heavy chains γ, α and δ have a constant region composed of *three* tandem (in a line) Ig domains, and a hinge region for added flexibility; heavy chains μ and ε have a constant region composed of *four* immunoglobulin domains. The variable region of the heavy chain differs in antibodies produced by different B cells, but is the same for all antibodies produced by a single B cell or B cell clone. The variable region of each heavy chain is approximately 110 amino acids long and is composed of a single Ig domain.

Light Chain

In mammals there are two types of immunoglobulin light chain, which are called lambda (λ) and kappa (κ). A light chain has two successive domains: one constant domain and one variable domain. The approximate length of a light chain is 211 to 217 amino acids. Each antibody contains two light chains that are always identical; only one type of light chain, κ or λ, is present per antibody in mammals. Other types of light chains, such as the iota (ι) chain, are found in other vertebrates like sharks (Chondrichthyes) and bony fishes (Teleostei).

CDRs, Fv, Fab and Fc Regions

Some parts of an antibody have the same functions. The arms of the Y, for example, contain the sites that can bind to antigens (in general, identical) and, therefore, recognize specific foreign objects. This region of the antibody is called the *Fab (fragment, antigen-binding) region*. It is composed of one constant and one variable domain from each heavy and light chain of the antibody. The paratope is shaped at the amino terminal end of the antibody monomer by the variable domains from the heavy and light chains. The variable domain is also referred to as the F_V region and is the most important region for binding to antigens. To be specific, variable loops of β-strands, three each on the light (V_L) and heavy (V_H) chains are responsible for binding to the antigen. These loops are referred to as the complementarity determining regions (CDRs). The structures of these CDRs have been clustered and classified by Chothia et al. and more recently by North et al. and Nikoloudis et al. In the framework of the immune network theory, CDRs are also called idiotypes. According to immune network theory, the adaptive immune system is regulated by interactions between idiotypes.

The base of the Y plays a role in modulating immune cell activity. This region is called the *Fc (Fragment, crystallizable) region*, and is composed of two heavy chains that contribute two or three constant domains depending on the class of the antibody. Thus, the Fc region ensures that each antibody generates an appropriate immune response for a given antigen, by binding to a specific class of Fc receptors, and other immune molecules, such as complement proteins. By doing this, it mediates different physiological effects including recognition of opsonized particles (binding to FcγR), lysis of cells (binding to complement), and degranulation of mast cells, basophils, and eosinophils (binding to FcεR).

In summary, the Fab region of the antibody determines antigen specificity while the Fc region of the antibody determines the antibody's class effect. Since only the constant domains of the heavy chains make up the Fc region of an antibody, the classes of heavy chain in antibodies determine their class effects. Possible classes of heavy chains in antibodies include alpha, gamma, delta, epsilon, and mu, and they define the antibody's isotypes IgA, G, D, E, and M, respectively. This infers different isotypes of antibodies have different class effects due to their different Fc regions binding and activating different types of receptors. Possible class effects of antibodies include: Opsonisation, agglutination, haemolysis, complement activation, mast cell degranulation, and neutralisation (though this class effect may be mediated by the Fab region rather than the Fc region). It also implies that Fab-mediated effects are directed at microbes or toxins, whilst Fc mediated effects are directed at effector cells or effector molecules.

Function

The main categories of antibody action include the following:

- Neutralisation, in which neutralizing antibodies block parts of the surface of a bacterial cell or virion to render its attack ineffective

- Agglutination, in which antibodies "glue together" foreign cells into clumps that are attractive targets for phagocytosis

- Precipitation, in which antibodies "glue together" serum-soluble antigens, forcing them to precipitate out of solution in clumps that are attractive targets for phagocytosis

- Complement activation (fixation), in which antibodies that are latched onto a foreign cell encourage complement to attack it with a membrane attack complex, which leads to the following:

  o Lysis of the foreign cell

  o Encouragement of inflammation by chemotactically attracting inflammatory cells

Activated B cells differentiate into either antibody-producing cells called plasma cells that secrete soluble antibody or memory cells that survive in the body for years afterward in order to allow the immune system to remember an antigen and respond faster upon future exposures.

At the prenatal and neonatal stages of life, the presence of antibodies is provided by passive immunization from the mother. Early endogenous antibody production varies for different kinds of antibodies, and usually appear within the first years of life. Since antibodies exist freely in the bloodstream, they are said to be part of the humoral immune system. Circulating antibodies are produced by clonal B cells that specifically respond to only one antigen (an example is a viruscapsid protein fragment). Antibodies contribute to immunity in three ways: They prevent pathogens from entering or damaging cells by binding to them; they stimulate removal of pathogens by macrophages and other cells by coating the pathogen; and they trigger destruction of pathogens by stimulating other immune responses such as the complement pathway. Antibodies will also trigger vasoactive amine degranulation to contribute to immunity against certain types of antigens (helminths, allergens).

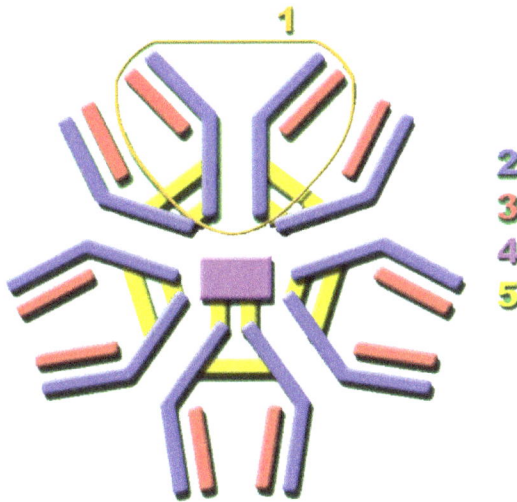

The secreted mammalian IgM has five Ig units. Each Ig unit (labeled 1) has two epitope binding Fab regions, so IgM is capable of binding up to 10 epitopes.

Activation of Complement

Antibodies that bind to surface antigens (for example, on bacteria) will attract the first component of the complement cascade with their Fc region and initiate activation of the "classical" complement system. This results in the killing of bacteria in two ways. First, the binding of the antibody and complement molecules marks the microbe for ingestion by phagocytes in a process called opsonization; these phagocytes are attracted by certain complement molecules generated in the complement cascade. Second, some complement system components form a membrane attack complex to assist antibodies to kill the bacterium directly (bacteriolysis).

Activation of Effector Cells

To combat pathogens that replicate outside cells, antibodies bind to pathogens to link them together, causing them to agglutinate. Since an antibody has at least two paratopes, it can bind more than one antigen by binding identical epitopes carried on the surfaces of these antigens. By coating the pathogen, antibodies stimulate effector functions against the pathogen in cells that recognize their Fc region.

Those cells that recognize coated pathogens have Fc receptors, which, as the name suggests, interact with the Fc region of IgA, IgG, and IgE antibodies. The engagement of a particular antibody with the Fc receptor on a particular cell triggers an effector function of that cell; phagocytes will phagocytose, mast cells and neutrophils will degranulate, natural killer cells will release cytokines and cytotoxic molecules; that will ultimately result in destruction of the invading microbe. The activation of natural killer cells by antibodies initiates a cytotoxic mechanism known as antibody-dependent cell-mediated cytotoxicity (ADCC) – this process may explain the efficacy of monoclonal antibodies used in biological therapies against cancer. The Fc receptors are isotype-specific, which gives greater flexibility to the immune system, invoking only the appropriate immune mechanisms for distinct pathogens.

Natural Antibodies

Humans and higher primates also produce "natural antibodies" that are present in serum before viral infection. Natural antibodies have been defined as antibodies that are produced without any previous infection, vaccination, other foreign antigen exposure or passive immunization. These antibodies can activate the classical complement pathway leading to lysis of enveloped virus particles long before the adaptive immune response is activated. Many natural antibodies are directed against the disaccharide galactose $\alpha(1,3)$-galactose (α-Gal), which is found as a terminal sugar on glycosylated cell surface proteins, and generated in response to production of this sugar by bacteria contained in the human gut. Rejection of xenotransplantated organs is thought to be, in part, the result of natural antibodies circulating in the serum of the recipient binding to α-Gal antigens expressed on the donor tissue.

Immunoglobulin Diversity

Virtually all microbes can trigger an antibody response. Successful recognition and eradication of many different types of microbes requires diversity among antibodies; their amino acid composition varies allowing them to interact with many different antigens. It has been estimated that humans generate about 10 billion different antibodies, each capable of binding a distinct epitope of an antigen. Although a huge repertoire of different antibodies is generated in a single individual, the number of genes available to make these proteins is limited by the size of the human genome. Several complex genetic mechanisms have evolved that allow vertebrate B cells to generate a diverse pool of antibodies from a relatively small number of antibody genes.

Domain Variability

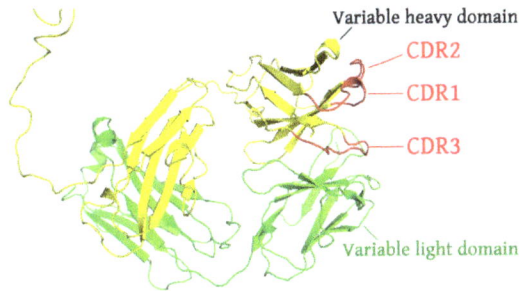

The complementarity determining regions of the heavy chain are shown in red (PDB: 1IGT)

The chromosomal region that encodes an antibody is large and contains several distinct gene loci for each domain of the antibody—the chromosome region containing heavy chain genes (IGH@) is found on chromosome 14, and the loci containing lambda and kappa light chain genes (IGL@ and IGK@) are found on chromosomes 22 and 2 in humans. One of these domains is called the variable domain, which is present in each heavy and light chain of every antibody, but can differ in different antibodies generated from distinct B cells. Differences, between the variable domains, are located on three loops known as hypervariable regions (HV-1, HV-2 and HV-3) or complementarity determining regions (CDR1, CDR2 and CDR3). CDRs are supported within the variable domains by conserved framework regions. The heavy chain locus contains about 65 different variable domain genes that all differ in their CDRs. Combining these genes with an array of genes for other domains of the antibody generates a large cavalry of antibodies with a high degree of variability. This combination is called V(D)J recombination discussed below.

V(D)J Recombination

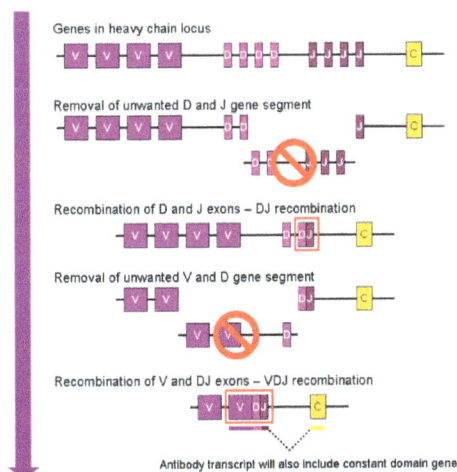

Simplified overview of V(D)J recombination of immunoglobulin heavy chains

Somatic recombination of immunoglobulins, also known as *V(D)J recombination*, involves the generation of a unique immunoglobulin variable region. The variable region of each immunoglobulin heavy or light chain is encoded in several pieces—known as gene segments (subgenes). These segments are called variable (V), diversity (D) and joining (J) segments. V, D and J segments are found in Ig heavy chains, but only V and J segments are found in Ig light chains. Multiple copies of the V, D and J gene segments exist, and are tandemly arranged in the genomes of mammals. In the bone marrow, each developing B cell will assemble an immunoglobulin variable region by randomly selecting and combining one V, one D and one J gene segment (or one V and one J segment in the light chain). As there are multiple copies of each type of gene segment, and different combinations of gene segments can be used to generate each immunoglobulin variable region, this process generates a huge number of antibodies, each with different paratopes, and thus different antigen specificities. Interestingly, the rearrangement of several subgenes (i.e. V2 family) for lambda light chain immunoglobulin is coupled with the activation of microRNA miR-650, which further influences biology of B-cells.

RAG proteins play an important role with V(D)J recombination in cutting DNA at a particular region. Without the presence of these proteins, V(D)J recombination would not occur.

After a B cell produces a functional immunoglobulin gene during V(D)J recombination, it cannot express any other variable region (a process known as allelic exclusion) thus each B cell can produce antibodies containing only one kind of variable chain.

Somatic Hypermutation and Affinity Maturation

Following activation with antigen, B cells begin to proliferate rapidly. In these rapidly dividing cells, the genes encoding the variable domains of the heavy and light chains undergo a high rate of point mutation, by a process called *somatic hypermutation* (SHM). SHM results in approximately one nucleotide change per variable gene, per cell division. As a consequence, any daughter B cells will acquire slight amino acid differences in the variable domains of their antibody chains.

This serves to increase the diversity of the antibody pool and impacts the antibody's antigen-binding affinity. Some point mutations will result in the production of antibodies that have a weaker interaction (low affinity) with their antigen than the original antibody, and some mutations will generate antibodies with a stronger interaction (high affinity). B cells that express high affinity antibodies on their surface will receive a strong survival signal during interactions with other cells, whereas those with low affinity antibodies will not, and will die by apoptosis. Thus, B cells expressing antibodies with a higher affinity for the antigen will outcompete those with weaker affinities for function and survival. The process of generating antibodies with increased binding affinities is called *affinity maturation*. Affinity maturation occurs in mature B cells after V(D)J recombination, and is dependent on help from helper T cells.

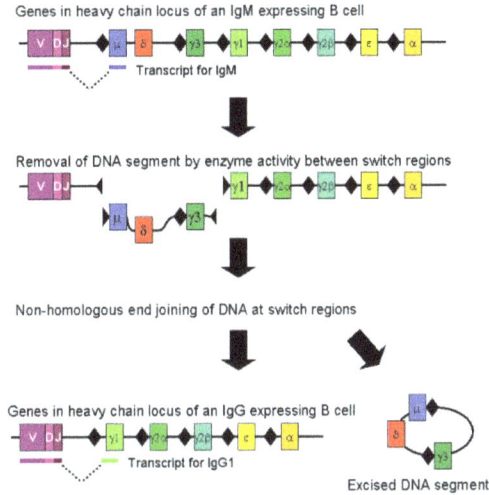

Mechanism of class switch recombination that allows isotype switching in activated B cells

Class Switching

Isotype or class switching is a biological process occurring after activation of the B cell, which allows the cell to produce different classes of antibody (IgA, IgE, or IgG). The different classes of antibody, and thus effector functions, are defined by the constant (C) regions of the immunoglobulin heavy chain. Initially, naive B cells express only cell-surface IgM and IgD with identical antigen binding regions. Each isotype is adapted for a distinct function; therefore, after activation, an antibody with an IgG, IgA, or IgE effector function might be required to effectively eliminate an antigen. Class switching allows different daughter cells from the same activated B cell to produce antibodies of different isotypes. Only the constant region of the antibody heavy chain changes during class switching; the variable regions, and therefore antigen specificity, remain unchanged. Thus the progeny of a single B cell can produce antibodies, all specific for the same antigen, but with the ability to produce the effector function appropriate for each antigenic challenge. Class switching is triggered by cytokines; the isotype generated depends on which cytokines are present in the B cell environment.

Class switching occurs in the heavy chain gene locus by a mechanism called class switch recombination (CSR). This mechanism relies on conserved nucleotide motifs, called *switch (S) regions*, found in DNA upstream of each constant region gene (except in the δ-chain). The DNA strand is broken by the activity of a series of enzymes at two selected S-regions. The variable domain exon is rejoined through a process called non-homologous end joining (NHEJ) to the desired constant region (γ, α or ε). This process results in an immunoglobulin gene that encodes an antibody of a different isotype.

Affinity Designations

A group of antibodies can be called *monovalent* (or *specific*) if they have affinity for the same epitope, or for the same antigen (but potentially different epitopes on the molecule), or for the same strain of microorganism (but potentially different antigens on or in it). In contrast, a group of antibodies can be called *polyvalent* (or *unspecific*) if they have affinity for various antigens or microorganisms.Intravenous immunoglobulin, if not otherwise noted, consists of polyvalent IgG. In contrast, monoclonal antibodies are monovalent for the same epitope.

Asymmetrical Antibodies

Heterodimeric antibodies, which are also asymmetrical and antibodies, allow for greater flexibility and new formats for attaching a variety of drugs to the antibody arms. One of the general formats for a heterodimeric antibody is the "knobs-into-holes" format. This format is specific to the heavy chain part of the constant region in antibodies. The "knobs" part is engineered by replacing a small amino acid with a larger one. It fits into the "hole", which is engineered by replacing a large amino acid with a smaller one. What connects the "knobs" to the "holes" are the disulfide bonds between each chain. The "knobs-into-holes" shape facilitates antibody dependent cell mediated cytotoxicity. Single chain variable fragments (scFv) are connected to the variable domain of the heavy and light chain via a short linker peptide. The linker is rich in glycine, which gives it more flexibility, and serine/threonine, which gives it specificity. Two different scFv fragments can be connected together, via a hinge region, to the constant domain of the heavy chain or the constant domain of the light chain. This gives the antibody bispecificity, allowing for the binding specificities of two different antigens. The "knobs-into-holes" format enhances heterodimer formation but doesn't suppress homodimer formation.

To further improve the function of heterodimeric antibodies, many scientists are looking towards artificial constructs. Artificial antibodies are largely diverse protein motifs that use the functional strategy of the antibody molecule, but aren't limited by the loop and framework structural constraints of the natural antibody. Being able to control the combinational design of the sequence and three-dimensional space could transcend the natural design and allow for the attachment of different combinations of drugs to the arms.

Heterodimeric antibodies have a greater range in shapes they can take and the drugs that are attached to the arms don't have to be the same on each arm, allowing for different combinations of drugs to be used in cancer treatment. Pharmaceuticals are able to produce highly functional bispecific, and even multispecific, antibodies. The degree to which they can function is impressive given that such a change shape from the natural form should lead to decreased functionality.

Medical Applications

Disease Diagnosis

Detection of particular antibodies is a very common form of medical diagnostics, and applications such as serology depend on these methods. For example, in biochemical assays for disease diagnosis, a titer of antibodies directed against Epstein-Barr virus or Lyme disease is estimated from the blood. If those antibodies are not present, either the person is not infected or the infection occurred a *very* long time ago, and the B cells generating these specific antibodies have naturally decayed.

In clinical immunology, levels of individual classes of immunoglobulins are measured by nephelometry (or turbidimetry) to characterize the antibody profile of patient. Elevations in different classes of immunoglobulins are sometimes useful in determining the cause of liver damage in patients for whom the diagnosis is unclear. For example, elevated IgA indicates alcoholic cirrhosis, elevated IgM indicates viral hepatitis and primary biliary cirrhosis, while IgG is elevated in viral hepatitis, autoimmune hepatitis and cirrhosis.

Autoimmune disorders can often be traced to antibodies that bind the body's own epitopes; many can be detected through blood tests. Antibodies directed against red blood cell surface antigens in immune mediated hemolytic anemia are detected with the Coombs test. The Coombs test is also used for antibody screening in blood transfusion preparation and also for antibody screening in antenatal women.

Practically, several immunodiagnostic methods based on detection of complex antigen-antibody are used to diagnose infectious diseases, for example ELISA, immunofluorescence, Western blot, immunodiffusion, immunoelectrophoresis, and magnetic immunoassay. Antibodies raised against human chorionic gonadotropin are used in over the counter pregnancy tests.

New dioxaborolane chemistry enables radioactive fluoride (^{18}F) labeling of antibodies, which allows for positron emission tomography (PET) imaging of cancer.

Disease Therapy

Targeted monoclonal antibody therapy is employed to treat diseases such as rheumatoid arthritis,multiple sclerosis,psoriasis, and many forms of cancer including non-Hodgkin's lymphoma,colorectal cancer, head and neck cancer and breast cancer.

Some immune deficiencies, such as X-linked agammaglobulinemia and hypogammaglobulinemia, result in partial or complete lack of antibodies. These diseases are often treated by inducing a short term form of immunity called passive immunity. Passive immunity is achieved through the transfer of ready-made antibodies in the form of human or animal serum, pooled immunoglobulin or monoclonal antibodies, into the affected individual.

Prenatal Therapy

Rhesus factor, also known as Rhesus D (RhD) antigen, is an antigen found on red blood cells; individuals that are Rhesus-positive (Rh+) have this antigen on their red blood cells and individuals that are Rhesus-negative (Rh−) do not. During normal childbirth, delivery trauma or complications during pregnancy, blood from a fetus can enter the mother's system. In the case of an Rh-incompatible mother and child, consequential blood mixing may sensitize an Rh- mother to the Rh antigen on the blood cells of the Rh+ child, putting the remainder of the pregnancy, and any subsequent pregnancies, at risk for hemolytic disease of the newborn.

Rho(D) immune globulin antibodies are specific for human Rhesus D (RhD) antigen. Anti-RhD antibodies are administered as part of a prenatal treatment regimen to prevent sensitization that may occur when a Rhesus-negative mother has a Rhesus-positive fetus. Treatment of a mother with Anti-RhD antibodies prior to and immediately after trauma and delivery destroys Rh antigen in the mother's system from the fetus. It is important to note that this occurs before the antigen can stimulate maternal B cells to "remember" Rh antigen by generating memory B cells. Therefore, her humoral immune system will not make anti-Rh antibodies, and will not attack the Rhesus antigens of the current or subsequent babies. Rho(D) Immune Globulin treatment prevents sensitization that can lead to Rh disease, but does not prevent or treat the underlying disease itself.

Research Applications

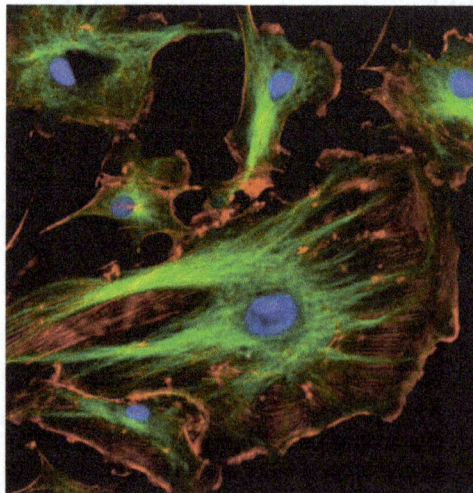

Immunofluorescence image of the eukaryotic cytoskeleton.
Actin filaments are shown in red, microtubules in green, and the nuclei in blue.

Specific antibodies are produced by injecting an antigen into a mammal, such as a mouse, rat, rabbit, goat, sheep, or horse for large quantities of antibody. Blood isolated from these animals contains *polyclonal antibodies*—multiple antibodies that bind to the same antigen—in the serum, which can now be called antiserum. Antigens are also injected into chickens for generation of polyclonal antibodies in egg yolk. To obtain antibody that is specific for a single epitope of an antigen, antibody-secreting lymphocytes are isolated from the animal and immortalized by fusing them with a cancer cell line. The fused cells are called hybridomas, and will continually grow and secrete antibody in culture. Single hybridoma cells are isolated by dilution cloning to generate cell clones that all produce the same antibody; these antibodies are called *monoclonal antibodies*. Polyclonal and monoclonal antibodies are often purified using Protein A/G or antigen-affinity chromatography.

In research, purified antibodies are used in many applications. Antibodies for research applications can be found directly from antibody suppliers, or through use of a specialist search engine. Research antibodies are most commonly used to identify and locate intracellular and extracellular proteins. Antibodies are used in flow cytometry to differentiate cell types by the proteins they express; different types of cell express different combinations of cluster of differentiation molecules on their surface, and produce different intracellular and secretable proteins. They are also used in immunoprecipitation to separate proteins and anything bound to them (co-immunoprecipitation) from other molecules in a cell lysate, in Western blot analyses to identify proteins separated by electrophoresis, and in immunohistochemistry or immunofluorescence to examine protein expression in tissue sections or to locate proteins within cells with the assistance of a microscope. Proteins can also be detected and quantified with antibodies, using ELISA and ELISPOT techniques.

- Reproducibility in Science

Antibodies used in research are some of the most powerful, yet most problematic reagents with a tremendous number of factors that must be controlled in any experiment including cross reactivity, or the antibody recognizing multiple epitopes and affinity, which can vary widely different depending on experimental conditions such as pH, solvent, state of tissue etc. Multiple attempts have been made to improve both the way that researchers validate antibodies and ways in which they report on antibodies. Researchers using antibodies in their work need to record them correctly in order to allow their research to be reproducible (and therefore tested, and qualified by other researchers). Less than half of research antibodies referenced in academic papers can be easily identified. Papers published in F1000 in 2014 and 2015 provide researchers with a guide for reporting research antibody use. The RRID paper, is co-published in 4 journals that implemented the RRIDs Standard for research resource citation, which draws data from the antibodyregistry.org as the source of antibody identifiers

Regulatory Validation of Monoclonal Antibody Products for Human Use

Production and testing :

Traditionally, most antibodies are produced by hybridoma cell lines through immortalization of antibody-producing cells by chemically-induced fusion with myeloma cells. In some cases, additional fusions with other lines have created "triomas" and "quadromas". The manufacturing process should be appropriately described and validated. Validation studies should at least include :

- The demonstration that the process is able to produce in good quality (the process should be validated)

- The efficiency of the antibody purification (all impurities and virus must be eliminated)

- The characterization of purified antibody (physicochemical characterization, immunological properties, biological activities, contaminants, ...)

- Determination of the virus clearance studies

Before clinical trials, studies of product safety and feasibility have to be performed :

- Product safety testing : Sterility (bacteria and fungi), In vitro and in vivo testing for adventitious viruses, Murine retrovirus testing... Product safety data needed before the initiation of feasibility trials in serious or immediately life-threatening conditions, it serves to evaluate dangerous potential of the product.

- Feasibility testing : These are pilot studies whose objectives include, among others, early characterization of safety and initial proof of concept in a small specific patient population (in vito or in vivo testing).

Preclinical studies :

Example of tissue cross-reactivity on human tissue microarray

- Testing cross-reactivity of antibody : to highlight unwanted interactions (toxicity) of antibodies with previously characterized tissues. This study can be performed in vitro (Reactivity of the antibody or immunoconjugate should be determined with a quick-frozen adult tissues) or in vivo (with appropriates animal models). More informations about in vitro cross-reactivity testing.

- Preclinical pharmacology and toxicity testing : Preclinical safety testing of antibody is designed to identify possible toxicities in humans, to estimate the likelihood and severity of potential adverse events in humans, and to identify a safe starting dose and dose escalation, when possible.

- Animal toxicity studies : Acute toxicity testing, Repeat-dose toxicity testing, Long-term toxicity testing http://www.animalresearch.info/en/drug-development/safety-testing/

- Pharmacokinetics and pharmacodynamics testing : Use for determinate clinical dosages, antibody activities (AUC, pharmacodynamics, biodistribution, ...), evaluation of the potential clinical effects

Structure Prediction

The importance of antibodies in health care and the biotechnology industry demands knowledge of their structures at high resolution. This information is used for protein engineering, modifying the antigen binding affinity, and identifying an epitope, of a given antibody. X-ray crystallography is one commonly used method for determining antibody structures. However, crystallizing an antibody is often laborious and time-consuming. Computational approaches provide a cheaper and faster alternative to crystallography, but their results are more equivocal, since they do not produce empirical structures. Online web servers such as *Web Antibody Modeling* (WAM) and *Prediction of Immunoglobulin Structure* (PIGS) enables computational modeling of antibody variable regions. Rosetta Antibody is a novel antibody F_v region structure prediction server, which incorporates sophisticated techniques to minimize CDR loops and optimize the relative orientation of the light and heavy chains, as well as homology models that predict successful docking of antibodies with their unique antigen.

The ability to describe the antibody through binding affinity to the antigen is supplemented by information on antibody structure and amino acid sequences for the purpose of patent claims.

History

The first use of the term "antibody" occurred in a text by Paul Ehrlich. The term *Antikörper* (the German word for *antibody*) appears in the conclusion of his article "Experimental Studies on Immunity", published in October 1891, which states that, "if two substances give rise to two different antikörper, then they themselves must be different". However, the term was not accepted immediately and several other terms for antibody were proposed; these included *Immunkörper, Amboceptor, Zwischenkörper, substance sensibilisatrice, copula, Desmon, philocytase, fixateur*, and *Immunisin*. The word *antibody* has formal analogy to the word *antitoxin* and a similar concept to *Immunkörper* (*immune body* in English). As such, the original construction of the word contains a logical flaw; the antitoxin is something directed against a toxin, while the antibody is a body directed against something.

Angel of the West (2008) by Julian Voss-Andreae is a sculpture based on the antibody structure published by E. Padlan. Created for the Florida campus of the Scripps Research Institute, the antibody is placed into a ring referencing Leonardo da Vinci's *Vitruvian Man* thus highlighting the similarity of the antibody and the human body.

The study of antibodies began in 1890 when Kitasato Shibasaburō described antibody activity against diphtheria and tetanus toxins. Kitasato put forward the theory of humoral immunity, proposing that a mediator in serum could react with a foreign antigen. His idea prompted Paul Ehrlich to propose the side-chain theory for antibody and antigen interaction in 1897, when he hypothesized that receptors (described as "side-chains") on the surface of cells could bind specifically to toxins – in a «lock-and-key» interaction – and that this binding reaction is the trigger for the production of antibodies. Other researchers believed that antibodies existed freely in the blood and, in 1904, Almroth Wright suggested that soluble antibodies coated bacteria to label them for phagocytosis and killing; a process that he named opsoninization.

Michael Heidelberger

In the 1920s, Michael Heidelberger and Oswald Avery observed that antigens could be precipitated by antibodies and went on to show that antibodies are made of protein. The biochemical properties of antigen-antibody-binding interactions were examined in more detail in the late 1930s by John Marrack. The next major advance was in the 1940s, when Linus Pauling confirmed the lock-and-key theory proposed by Ehrlich by showing that the interactions between antibodies and antigens depend more on their shape than their chemical composition. In 1948, Astrid Fagreaus discovered that B cells, in the form of plasma cells, were responsible for generating antibodies.

Further work concentrated on characterizing the structures of the antibody proteins. A major advance in these structural studies was the discovery in the early 1960s by Gerald Edelman and Joseph Gally of the antibody light chain, and their realization that this protein is the same as the Bence-Jones protein described in 1845 by Henry Bence Jones. Edelman went on to discover that antibodies are composed of disulfide bond-linked heavy and light chains. Around the same time, antibody-binding (Fab) and antibody tail (Fc) regions of IgG were characterized by Rodney Porter. Together, these scientists deduced the structure and complete amino acid sequence of IgG, a feat for which they were jointly awarded the 1972 Nobel Prize in Physiology or Medicine. The Fv fragment was prepared and characterized by David Givol. While most of these early studies focused on IgM and IgG, other immunoglobulin isotypes were identified in the 1960s: Thomas Tomasi discovered secretory antibody (IgA); David S. Rowe and John L. Fahey discovered IgD; and Kimishige Ishizaka and Teruko Ishizaka discovered IgE and showed it was a class of antibodies involved in allergic reactions. In a landmark series of experiments beginning in 1976, Susumu Tonegawa showed that genetic material can rearrange itself to form the vast array of available antibodies.

Antibody Mimetic

Antibody mimetics are organic compounds that, like antibodies, can specifically bind antigens. They are usually artificial peptides or proteins with a molar mass of about 3 to 20 kDa. Nucleic acids and small molecules are sometimes considered antibody mimetics, but not artificial antibodies, antibody fragments and fusion proteins are composed from these. Common advantages over antibodies are better solubility, tissue penetration, stability towards heat and enzymes, and comparatively low production costs. Antibody mimetics such as the Affimer and the DARPin have being developed and commercialised as research, diagnostic and therapeutic agents.

Collagen

Collagen is the main structural protein in the extracellular space in the various connective tissues in animal bodies. As the main component of connective tissue, it is the most abundant protein in mammals, making up from 25% to 35% of the whole-body protein content. Depending upon the degree of mineralization, collagen tissues may be rigid (bone), compliant (tendon), or have a gradient from rigid to compliant (cartilage). Collagen, in the form of elongated fibrils, is mostly found in fibrous tissues such as tendons, ligaments and skin. It is also abundant in corneas, cartilage, bones, blood vessels, the gut, intervertebral discs and the dentin in teeth. In muscle tissue, it serves as a major component of the endomysium. Collagen constitutes one to two percent of muscle tissue, and accounts for 6% of the weight of strong, tendinous muscles. The fibroblast is the most common cell that creates collagen.

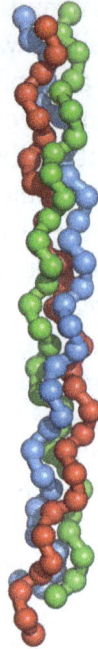

Tropocollagen molecule: three left-handed procollagens (red, green, blue) join to form a right handed triple helical tropocollagen.

Gelatin, which is used in food and industry, is collagen that has been irreversibly hydrolyzed. Collagen also has many medical uses in treating complications of the bones and skin.

Types of Collagen

Collagen occurs in many places throughout the body. Over 90% of the collagen in the human body, however, is type I.

So far, 28 types of collagen have been identified and described. They can be divided into several groups according to the structure they form:

- Fibrillar (Type I, II, III, V, XI)

- Non-fibrillar

- o FACIT (Fibril Associated Collagens with Interrupted Triple Helices) (Type IX, XII, XIV, XVI, XIX)

- o Short chain (Type VIII, X)

- o Basement membrane (Type IV)

- o Multiplexin (Multiple Triple Helix domains with Interruptions) (Type XV, XVIII)

- o MACIT (Membrane Associated Collagens with Interrupted Triple Helices) (Type XIII, XVII)

- o Other (Type VI, VII)

The five most common types are:

- Type I: skin, tendon, vascular ligature, organs, bone (main component of the organic part of bone)

- Type II: cartilage (main collagenous component of cartilage)

- Type III: reticulate (main component of reticular fibers), commonly found alongside type I.

- Type IV: forms basal lamina, the epithelium-secreted layer of the basement membrane.

- Type V: cell surfaces, hair and placenta

Medical uses

Cardiac Applications

The collagenous cardiac skeleton which includes the four heart valve rings, is histologically, elastically and uniquely bound to cardiac muscle. The cardiac skeleton also includes the separating septa of the heart chambers – the interventricular septum and the atrioventricular septum. Collagen contribution to the measure of cardiac performance summarily represents a continuous torsional force opposed to the fluid mechanics of blood pressure emitted from the heart. The collagenous structure that divides the upper chambers of the heart from the lower chambers is an impermeable membrane that excludes both blood and electrical impulses through typical physiological means. With support from collagen, atrial fibrillation should never deteriorate to ventricular fibrillation. Collagen is layered in variable densities with cardiac muscle mass. The mass, distribution, age and density of collagen all contribute to the compliance required to move blood back and forth. Individual cardiac valvular leaflets are folded into shape by specialized collagen under variable pressure. Gradual calcium deposition within collagen occurs as a natural function of aging. Calcified points within collagen matrices show contrast in a moving display of blood and muscle, enabling methods of cardiac imaging technology to arrive at ratios essentially stating blood in (cardiac input) and blood out (cardiac output). Pathology of the collagen underpinning of the heart is understood within the category of connective tissue disease.

Cosmetic Surgery

Collagen has been widely used in cosmetic surgery, as a healing aid for burn patients for reconstruction of bone and a wide variety of dental, orthopedic, and surgical purposes. Both human and bovine collagen is widely used as dermal fillers for treatment of wrinkles and skin aging. Some points of interest are:

1. When used cosmetically, there is a chance of allergic reactions causing prolonged redness; however, this can be virtually eliminated by simple and inconspicuous patch testing prior to cosmetic use.

2. Most medical collagen is derived from young beef cattle (bovine) from certified BSE-free animals. Most manufacturers use donor animals from either "closed herds", or from countries which have never had a reported case of BSE such as Australia, Brazil, and New Zealand.

Bone Grafts

As the skeleton forms the structure of the body, it is vital that it maintains its strength, even after breaks and injuries. Collagen is used in bone grafting as it has a triple helical structure, making it a very strong molecule. It is ideal for use in bones, as it does not compromise the structural integrity of the skeleton. The triple helical structure of collagen prevents it from being broken down by enzymes, it enables adhesiveness of cells and it is important for the proper assembly of the extracellular matrix.

Tissue Regeneration

Collagen scaffolds are used in tissue regeneration, whether in sponges, thin sheets, or gels. Collagen has the correct properties for tissue regeneration such as pore structure, permeability, hydrophilicity and it is stable in vivo. Collagen scaffolds are also ideal for the deposition of cells, such as osteoblasts and fibroblasts and once inserted, growth is able to continue as normal in the tissue.

Reconstructive Surgical uses

Collagens are widely employed in the construction of the artificial skin substitutes used in the management of severe burns. These collagens may be derived from bovine, equine, porcine, or even human sources; and are sometimes used in combination with silicones, glycosaminoglycans, fibroblasts, growth factors and other substances.

Collagen is also sold commercially in pill form as a supplement to aid joint mobility. However, because proteins are broken down into amino acids before absorption, there is no reason for orally ingested collagen to affect connective tissue in the body, except through the effect of individual amino acid supplementation.

Collagen is also frequently used in scientific research applications for cell culture, studying cell behavior and cellular interactions with the extracellular environment.

Wound Care

Collagen is one of the body's key natural resources and a component of skin tissue that can benefit all stages of the wound healing process. When collagen is made available to the wound bed, closure can occur. Wound deterioration, followed sometimes by procedures such as amputation, can thus be avoided.

Collagen is a natural product, therefore it is used as a natural wound dressing and has properties that artificial wound dressings do not have. It is resistant against bacteria, which is of vital importance in a wound dressing. It helps to keep the wound sterile, because of its natural ability to fight infection. When collagen is used as a burn dressing, healthy granulation tissue is able to form very quickly over the burn, helping it to heal rapidly.

Throughout the 4 phases of wound healing, collagen performs the following functions in wound healing:

- Guiding function: Collagen fibers serve to guide fibroblasts. Fibroblasts migrate along a connective tissue matrix.

- Chemotactic properties: The large surface area available on collagen fibers can attract fibrogenic cells which help in healing.

- Nucleation: Collagen, in the presence of certain neutral salt molecules can act as a nucleating agent causing formation of fibrillar structures. A collagen wound dressing might serve as a guide for orienting new collagen deposition and capillary growth.

- Hemostatic properties: Blood platelets interact with the collagen to make a hemostatic plug.

Veterinary Use

Some studies have shown efficacy of collagen supplementation for dogs with osteoarthritis pain, alone or in combination with other nutraceuticals like glucosamine and chondroitin.

Chemistry

The collagen protein is composed of a triple helix, which generally consists of two identical chains (α1) and an additional chain that differs slightly in its chemical composition (α2). The amino acid composition of collagen is atypical for proteins, particularly with respect to its high hydroxyproline content. The most common motifs in the amino acid sequence of collagen are glycine-proline-X and glycine-X-hydroxyproline, where X is any amino acid other than glycine, proline or hydroxyproline. The average amino acid composition for fish and mammal skin is given.

Amino acid	Abundance in mammal skin (residues/1000)	Abundance in fish skin (residues/1000)
Glycine	329	339
Proline	126	108
Alanine	109	114
Hydroxyproline	95	67
Glutamic acid	74	76
Arginine	49	52
Aspartic acid	47	47
Serine	36	46
Lysine	29	26
Leucine	24	23
Valine	22	21
Threonine	19	26
Phenylalanine	13	14
Isoleucine	11	11
Hydroxylysine	6	8
Methionine	6	13
Histidine	5	7
Tyrosine	3	3
Cysteine	1	1
Tryptophan	0	0

Synthesis

First, a three-dimensional stranded structure is assembled, with the amino acids glycine and proline as its principal components. This is not yet collagen but its precursor, procollagen. Procollagen is then modified by the addition of hydroxyl groups to the amino acids proline and lysine. This step is important for later glycosylation and the formation of the triple helix structure of collagen. The hydroxylase enzymes that perform these reactions require Vitamin C as a cofactor, and a deficiency in this vitamin results in impaired collagen synthesis and the resulting disease scurvy These hydroxylation reactions are catalyzed by two different enzymes: prolyl-4-hydroxylase and lysyl-hydroxylase. Vitamin C also serves with them in inducing these reactions. In this service, one molecule of vitamin C is destroyed for each H replaced by OH. The synthesis of collagen occurs inside and outside of the cell. The formation of collagen which results in fibrillary collagen (most common form) is discussed here. Meshwork collagen, which is often involved in the formation of filtration systems, is the other form of collagen. All types of collagens are triple helices, and the differences lie in the make-up of the alpha peptides created in step 2.

1. Transcription of mRNA: About 34 genes are associated with collagen formation, each coding for a specific mRNA sequence, and typically have the "*COL*" prefix. The beginning of collagen synthesis begins with turning on genes which are associated with the formation of a particular alpha peptide (typically alpha 1, 2 or 3).

2. Pre-pro-peptide formation: Once the final mRNA exits from the cell nucleus and enters into the cytoplasm, it links with the ribosomal subunits and the process of translation occurs. The early/first part of the new peptide is known as the signal sequence. The signal sequence on the N-terminal of the peptide is recognized by a signal recognition particle on the endoplasmic reticulum, which will be responsible for directing the pre-pro-peptide into the endoplasmic reticulum. Therefore, once the synthesis of new peptide is finished, it goes directly into the endoplasmic reticulum for post-translational processing. It is now known as pre-pro-collagen.

3. Pre-pro-peptide to pro-collagen: Three modifications of the pre-pro-peptide occur leading to the formation of the alpha peptide:

   1. The signal peptide on the N-terminal is dissolved, and the molecule is now known as *propeptide* (not procollagen).

   2. Hydroxylation of lysines and prolines on propeptide by the enzymes 'prolyl hydroxylase' and 'lysyl hydroxylase' (to produce hydroxyproline and hydroxylysine) occurs to aid cross-linking of the alpha peptides. This enzymatic step requires vitamin C as a cofactor. In scurvy, the lack of hydroxylation of prolines and lysines causes a looser triple helix (which is formed by three alpha peptides).

   3. Glycosylation occurs by adding either glucose or galactose monomers onto the hydroxyl groups that were placed onto lysines, but not on prolines.

   4. Once these modifications have taken place, three of the hydroxylated and glycosylated propeptides twist into a triple helix forming procollagen. Procollagen still has unwound ends, which will be later trimmed. At this point, the procollagen is packaged into a transfer vesicle destined for the Golgi apparatus.

4. Golgi apparatus modification: In the Golgi apparatus, the procollagen goes through one last post-translational modification before being secreted out of the cell. In this step, oligosaccharides (not monosaccharides as in step 3) are added, and then the procollagen is packaged into a secretory vesicle destined for the extracellular space.

5. Formation of tropocollagen: Once outside the cell, membrane bound enzymes known as 'collagen peptidases', remove the "loose ends" of the procollagen molecule. What is left is known as tropocollagen. Defects in this step produce one of the many collagenopathies known as Ehlers-Danlos syndrome. This step is absent when synthesizing type III, a type of fibrilar collagen.

6. Formation of the collagen fibril: 'Lysyl oxidase', an extracellular enzyme, produces the final step in the collagen synthesis pathway. This enzyme acts on lysines and hydroxylysines producing aldehyde groups, which will eventually undergo covalent bonding between tropocollagen molecules. This polymer of tropocollogen is known as a collagen fibril.

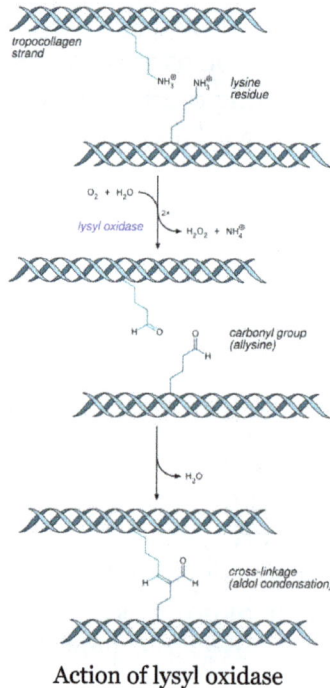

Action of lysyl oxidase

Amino Acids

Collagen has an unusual amino acid composition and sequence:

- Glycine is found at almost every third residue.

- Proline makes up about 17% of collagen.

- Collagen contains two uncommon derivative amino acids not directly inserted during translation. These amino acids are found at specific locations relative to glycine and are modified post-translationally by different enzymes, both of which require vitamin C as a cofactor.

    o   Hydroxyproline derived from proline

    o   Hydroxylysine derived from lysine - depending on the type of collagen, varying numbers of hydroxylysines are glycosylated (mostly having di-saccharides attached).

Cortisol stimulates degradation of (skin) collagen into amino acids.

Collagen I Formation

Most collagen forms in a similar manner, but the following process is typical for type I:

1. Inside the cell

    1. Two types of alpha chains are formed during translation on ribosomes along the rough endoplasmic reticulum (RER): alpha-1 and alpha-2 chains. These peptide chains (known as preprocollagen) have registration peptides on each end and a signal peptide.

    2. Polypeptide chains are released into the lumen of the RER.

    3. Signal peptides are cleaved inside the RER and the chains are now known as pro-alpha chains.

    4. Hydroxylation of lysine and proline amino acids occurs inside the lumen. This process is dependent on ascorbic acid (vitamin C) as a cofactor.

    5. Glycosylation of specific hydroxylysine residues occurs.

    6. Triple alpha helical structure is formed inside the endoplasmic reticulum from two alpha-1 chains and one alpha-2 chain.

    7. Procollagen is shipped to the Golgi apparatus, where it is packaged and secreted by exocytosis.

2. Outside the cell

    1. Registration peptides are cleaved and tropocollagen is formed by procollagen peptidase.

    2. Multiple tropocollagen molecules form collagen fibrils, via covalent cross-linking (aldol reaction) by lysyl oxidase which links hydroxylysine and lysine residues. Multiple collagen fibrils form into collagen fibers.

3. Collagen may be attached to cell membranes via several types of protein, including fibronectin and integrin.

Synthetic Pathogenesis

Vitamin C deficiency causes scurvy, a serious and painful disease in which defective collagen prevents the formation of strong connective tissue. Gums deteriorate and bleed, with loss of teeth; skin discolors, and wounds do not heal. Prior to the 18th century, this condition was notorious among long-duration military, particularly naval, expeditions during which participants were deprived of foods containing vitamin C.

An autoimmune disease such as lupus erythematosus or rheumatoid arthritis may attack healthy collagen fibers.

Many bacteria and viruses secrete virulence factors, such as the enzyme collagenase, which destroys collagen or interferes with its production.

Molecular Structure

A single collagen molecule, tropocollagen, is used to make up larger collagen aggregates, such as fibrils. It is approximately 300 nm long and 1.5 nm in diameter, and it is made up of three polypeptide strands (called alpha peptides, see step 2), each of which has the conformation of a left-handed helix – this should not be confused with the right-handed alpha helix. These three left-handed helices are twisted together into a right-handed triple helix or "super helix", a cooperative quaternary structure stabilized by many hydrogen bonds. With type I collagen and possibly all fibrillar collagens, if not all collagens, each triple-helix associates into a right-handed super-super-coil referred to as the collagen microfibril. Each microfibril is interdigitated with its neighboring microfibrils to a degree that might suggest they are individually unstable, although within collagen fibrils, they are so well ordered as to be crystalline.

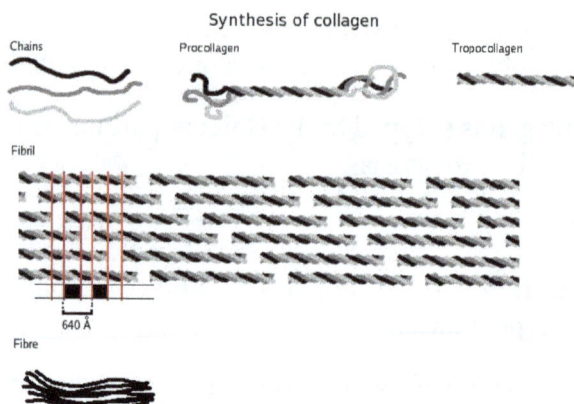

Three polypeptides coil to form tropocollagen. Many tropocollagens then bind together to form a fibril, and many of these then form a fibre.

A distinctive feature of collagen is the regular arrangement of amino acids in each of the three chains of these collagen subunits. The sequence often follows the pattern Gly-Pro-X or Gly-X-Hyp, where X may be any of various other amino acid residues. Proline or hydroxyproline constitute about 1/6 of the total sequence. With glycine accounting for the 1/3 of the sequence, this means approximately half of the collagen sequence is not glycine, proline or hydroxyproline, a fact often missed due to the distraction of the unusual GX_1X_2 character of collagen alpha-peptides. The high glycine content of collagen is important with respect to stabilization of the collagen helix as this allows the very close association of the collagen fibers within the molecule, facilitating hydrogen bonding and the formation of intermolecular cross-links. This kind of regular repetition and high glycine content is found in only a few other fibrous proteins, such as silkfibroin.

Collagen is not only a structural protein. Due to its key role in the determination of cell phenotype, cell adhesion, tissue regulation and infrastructure, many sections of its non-proline-rich regions have cell or matrix association / regulation roles. The relatively high content of proline and hydroxyproline rings, with their geometrically constrained carboxyl and (secondary) amino groups, along with the rich abundance of glycine, accounts for the tendency of the individual polypeptide strands to form left-handed helices spontaneously, without any intrachain hydrogen bonding.

Because glycine is the smallest amino acid with no side chain, it plays a unique role in fibrous structural proteins. In collagen, Gly is required at every third position because the assembly of the triple helix puts this residue at the interior (axis) of the helix, where there is no space for a larger side group than glycine's single hydrogen atom. For the same reason, the rings of the Pro and Hyp must point outward. These two amino acids help stabilize the triple helix—Hyp even more so than Pro; a lower concentration of them is required in animals such as fish, whose body temperatures are lower than most warm-blooded animals. Lower proline and hydroxyproline contents are characteristic of cold-water, but not warm-water fish; the latter tend to have similar proline and hydroxyproline contents to mammals. The lower proline and hydroxproline contents of cold-water fish and other poikilotherm animals leads to their collagen having a lower thermal stability than mammalian collagen. This lower thermal stability means that gelatin derived from fish collagen is not suitable for many food and industrial applications.

The tropocollagen subunits spontaneously self-assemble, with regularly staggered ends, into even larger arrays in the extracellular spaces of tissues. Additional assembly of fibrils is guided by fibroblasts, which deposit fully formed fibrils from fibripositors. In the fibrillar collagens, the molecules are staggered from each other by about 67 nm (a unit that is referred to as 'D' and changes depending upon the hydration state of the aggregate). Each D-period contains four plus a fraction collagen molecules, because 300 nm divided by 67 nm does not give an integer (the length of the collagen molecule divided by the stagger distance D). Therefore, in each D-period repeat of the microfibril, there is a part containing five molecules in cross-section, called the "overlap", and a part containing only four molecules, called the "gap". The triple-helices are also arranged in a hexagonal or quasihexagonal array in cross-section, in both the gap and overlap regions.

There is some covalent crosslinking within the triple helices, and a variable amount of covalent crosslinking between tropocollagen helices forming well organized aggregates (such as fibrils). Larger fibrillar bundles are formed with the aid of several different classes of proteins (including different collagen types), glycoproteins and proteoglycans to form the different types of mature tissues from alternate combinations of the same key players. Collagen's insolubility was a barrier to the study of monomeric collagen until it was found that tropocollagen from young animals can be extracted because it is not yet fully crosslinked. However, advances in microscopy techniques (i.e. electron microscopy (EM) and atomic force microscopy (AFM)) and X-ray diffraction have enabled researchers to obtain increasingly detailed images of collagen structure *in situ*. These later advances are particularly important to better understanding the way in which collagen structure affects cell–cell and cell–matrix communication, and how tissues are constructed in growth and repair, and changed in development and disease. For example, using AFM–based nanoindentation it has been shown that a single collagen fibril is a heterogeneous material along its axial direction with significantly different mechanical properties in its gap and overlap regions, correlating with its different molecular organizations in these two regions.

Collagen fibrils/aggregates are arranged in different combinations and concentrations in various tissues to provide varying tissue properties. In bone, entire collagen triple helices lie in a parallel, staggered array. 40 nm gaps between the ends of the tropocollagen subunits (approximately equal to the gap region) probably serve as nucleation sites for the deposition of long, hard, fine crystals of the mineral component, which is (approximately) $Ca_{10}(OH)_2(PO_4)_6$. Type I collagen gives bone its tensile strength.

Associated Disorders

Collagen-related diseases most commonly arise from genetic defects or nutritional deficiencies that affect the biosynthesis, assembly, postranslational modification, secretion, or other processes involved in normal collagen production.

Genetic Defects of Collagen Genes			
Type	**Notes**	**Gene(s)**	**Disorders**
I	This is the most abundant collagen of the human body. It is present in scar tissue, the end product when tissue heals by repair. It is found in tendons, skin, artery walls, cornea, the endomysium surrounding muscle fibers, fibrocartilage, and the organic part of bones and teeth.	COL1A1, COL1A2	Osteogenesis imperfecta, Ehlers–Danlos syndrome, Infantile cortical hyperostosis a.k.a. Caffey's disease

II	Hyaline cartilage, makes up 50% of all cartilage protein. Vitreous humour of the eye.	COL2A1	Collagenopathy, types II and XI
III	This is the collagen of granulation tissue, and is produced quickly by young fibroblasts before the tougher type I collagen is synthesized. Reticular fiber. Also found in artery walls, skin, intestines and the uterus	COL3A1	Ehlers–Danlos syndrome, Dupuytren's contracture
IV	Basal lamina; eye lens. Also serves as part of the filtration system in capillaries and the glomeruli of nephron in the kidney.	COL4A1, COL4A2, COL4A3, COL4A4, COL4A5, COL4A6	Alport syndrome, Goodpasture's syndrome
V	Most interstitial tissue, assoc. with type I, associated with placenta	COL5A1, COL5A2, COL5A3	Ehlers–Danlos syndrome (Classical)
VI	Most interstitial tissue, assoc. with type I	COL6A1, COL6A2, COL6A3, COL6A5	Ulrich myopathy, Bethlem myopathy, Atopic dermatitis
VII	Forms anchoring fibrils in dermoepidermal junctions	COL7A1	Epidermolysis bullosa dystrophica
VIII	Some endothelial cells	COL8A1, COL8A2	Posterior polymorphous corneal dystrophy 2
IX	FACIT collagen, cartilage, assoc. with type II and XI fibrils	COL9A1, COL9A2, COL9A3	EDM2 and EDM3
X	Hypertrophic and mineralizing cartilage	COL10A1	Schmid metaphyseal dysplasia
XI	Cartilage	COL11A1, COL11A2	Collagenopathy, types II and XI
XII	FACIT collagen, interacts with type I containing fibrils, decorin and glycosaminoglycans	COL12A1	–
XIII	Transmembrane collagen, interacts with integrin a1b1, fibronectin and components of basement membranes like nidogen and perlecan.	COL13A1	–
XIV	FACIT collagen, also known as undulin	COL14A1	–
XV	–	COL15A1	–
XVI	–	COL16A1	–
XVII	Transmembrane collagen, also known as BP180, a 180 kDa protein	COL17A1	Bullous pemphigoid and certain forms of junctional epidermolysis bullosa
XVIII	Source of endostatin	COL18A1	–
XIX	FACIT collagen	COL19A1	–
XX	–	COL20A1	–
XXI	FACIT collagen	COL21A1	–
XXII	–	COL22A1	–
XXIII	MACIT collagen	COL23A1	–
XXIV	–	COL24A1	–
XXV	–	COL25A1	–
XXVI	–	EMID2	–

| XXVII | – | COL27A1 | – |
| XXVIII | – | COL28A1 | – |

In addition to the above-mentioned disorders, excessive deposition of collagen occurs in scleroderma.

Diseases

One thousand mutations have been identified in 12 out of more than 20 types of collagen. These mutations can lead to various diseases at the tissue level.

Osteogenesis imperfecta – Caused by a mutation in type 1 collagen, dominant autosomal disorder, results in weak bones and irregular connective tissue, some cases can be mild while others can be lethal, mild cases have lowered levels of collagen type 1 while severe cases have structural defects in collagen.

Chondrodysplasias – Skeletal disorder believed to be caused by a mutation in type 2 collagen, further research is being conducted to confirm this.

Ehlers-Danlos syndrome – Six different types of this disorder, which lead to deformities in connective tissue, are known. Some types can be lethal, leading to the rupture of arteries. Each syndrome is caused by a different mutation, for example type four of this disorder is caused by a mutation in collagen type 3.

Alport syndrome – Can be passed on genetically, usually as X-linked dominant, but also as both an autosomal dominant and autosomal recessive disorder, sufferers have problems with their kidneys and eyes, loss of hearing can also develop in during the childhood or adolescent years.

Osteoporosis – Not inherited genetically, brought on with age, associated with reduced levels of collagen in the skin and bones, growth hormone injections are being researched as a possible treatment to counteract any loss of collagen.

Knobloch syndrome – Caused by a mutation in the COL18A1 gene that codes for the production of collagen XVIII. Patients present with protrusion of the brain tissue and degeneration of the retina, an individual who has family members suffering from the disorder are at an increased risk of developing it themselves as there is a hereditary link.

Characteristics

Collagen is one of the long, fibrous structural proteins whose functions are quite different from those of globular proteins, such as enzymes. Tough bundles of collagen called *collagen fibers* are a major component of the extracellular matrix that supports most tissues and gives cells structure from the outside, but collagen is also found inside certain cells. Collagen has great tensile strength, and is the main component of fascia, cartilage, ligaments, tendons, bone and skin. Along with elastin and soft keratin, it is responsible for skin strength and elasticity, and its degradation leads to wrinkles that accompany aging. It strengthens blood vessels and plays a role in tissue development. It is present in the cornea and lens of the eye in crystalline form. It may be one of the most abundant proteins in the fossil record, given that it appears to fossilize frequently, even in bones from the Mesozoic and Paleozoic.

Uses

Collagen has a wide variety of applications, from food to medical. For instance, it is used in cosmetic surgery and burn surgery. It is widely used in the form of collagen casings for sausages, which are also used in the manufacture of musical strings.

If collagen is subject to sufficient denaturation, e.g. by heating, the three tropocollagen strands separate partially or completely into globular domains, containing a different secondary structure to the normal collagen polyproline II (PPII), e.g. random coils. This process describes the formation of gelatin, which is used in many foods, including flavored gelatin desserts. Besides food, gelatin has been used in pharmaceutical, cosmetic, and photography industries.

Collagen adhesive was used by Egyptians about 4,000 years ago, and Native Americans used it in bows about 1,500 years ago. The oldest glue in the world, carbon-dated as more than 8,000 years old, was found to be collagen—used as a protective lining on rope baskets and embroideredfabrics, and to hold utensils together; also in crisscross decorations on human skulls. Collagen normally converts to gelatin, but survived due to dry conditions. Animal glues are thermoplastic, softening again upon reheating, so they are still used in making musical instruments such as fine violins and guitars, which may have to be reopened for repairs—an application incompatible with tough, syntheticplastic adhesives, which are permanent. Animal sinews and skins, including leather, have been used to make useful articles for millennia.

Gelatin-resorcinol-formaldehyde glue (and with formaldehyde replaced by less-toxic pentanedial and ethanedial) has been used to repair experimental incisions in rabbitlungs.

History

The molecular and packing structures of collagen have eluded scientists over decades of research. The first evidence that it possesses a regular structure at the molecular level was presented in the mid-1930s. Since that time, many prominent scholars, including Nobel laureates Crick, Pauling, Rich and Yonath, and others, including Brodsky, Berman, and Ramachandran, concentrated on the conformation of the collagen monomer. Several competing models, although correctly dealing with the conformation of each individual peptide chain, gave way to the triple-helical "Madras" model of Ramachandran, which provided an essentially correct model of the molecule's quaternary structure although this model still required some refinement. The packing structure of collagen has not been defined to the same degree outside of the fibrillar collagen types, although it has been long known to be hexagonal or quasi-hexagonal. As with its monomeric structure, several conflicting models alleged that either the packing arrangement of collagen molecules is 'sheet-like' or microfibrillar. The microfibrillar structure of collagen fibrils in tendon, cornea and cartilage has been directly imaged by electron microscopy. The microfibrillar structure of tail tendon, as described by Fraser, Miller, and Wess (amongst others), was modeled as being closest to the observed structure, although it oversimplified the topological progression of neighboring collagen molecules, and hence did not predict the correct conformation of the discontinuous D-periodic pentameric arrangement termed simply: the microfibril. Various cross linking agents like L-Dopaquinone, embeline, potassium embelate and 5-O-methyl embelin could be developed as potential cross-linking/stabilization agents of collagen preparation and its application as wound dressing sheet in clinical applications is enhanced.

Transfer RNA

A transfer RNA (abbreviated tRNA and formerly referred to as sRNA, for soluble RNA) is an adaptor molecule composed of RNA, typically 76 to 90 nucleotides in length, that serves as the physical link between the mRNA and the amino acid sequence of proteins. It does this by carrying an amino acid to the protein synthetic machinery of a cell (ribosome) as directed by a three-nucleotide sequence (codon) in a messenger RNA (mRNA). As such, tRNAs are a necessary component of translation, the biological synthesis of new proteins in accordance with the genetic code.

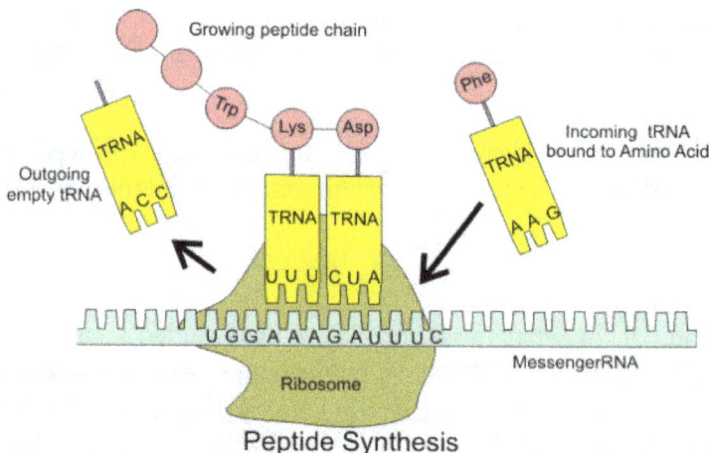

Peptide Synthesis

The interaction of tRNA and mRNA in protein synthesis.

Overview

While the specific nucleotide sequence of an mRNA specifies which amino acids are incorporated into the protein product of the gene from which the mRNA is transcribed, the role of *tRNA* is to specify which sequence from the genetic code corresponds to which amino acid. The mRNA encodes a protein as a series of contiguous codons, each of which is recognized by a particular tRNA. One end of the tRNA matches the genetic code in a three-nucleotide sequence called the anticodon. The anticodon forms three base pairs with a codon in mRNA during protein biosynthesis. On the other end of the tRNA is a covalent attachment to the amino acid that corresponds to the anticodon sequence. Each type of tRNA molecule can be attached to only one type of amino acid, so each organism has many types of tRNA. Because the genetic code contains multiple codons that specify the same amino acid, there are several tRNA molecules bearing different anticodons which carry the same amino acid.

The covalent attachment to the tRNA 3' end is catalyzed by enzymes called amino-acyl tRNA synthetases. During protein synthesis, tRNAs with attached amino acids are delivered to the ribosome by proteins called elongation factors, which aid in association of the tRNA with the ribosome, synthesis of the new polypeptide and translocation (movement) of the ribosome along the mRNA. If the tRNA's anticodon matches the mRNA, another tRNA already bound to the ribosome transfers the growing polypeptide chain from its 3' end to the amino acid attached to the 3' end of the newly delivered tRNA, a reaction catalyzed by the ribosome. A large number of the individual nucleotides in a tRNA molecule may be chemically modified, often by methylation or deamidation. These unusual bases sometimes affect the tRNA's interaction with ribosomes and sometimes occur in the anticodon to alter base-pairing properties.

Structure

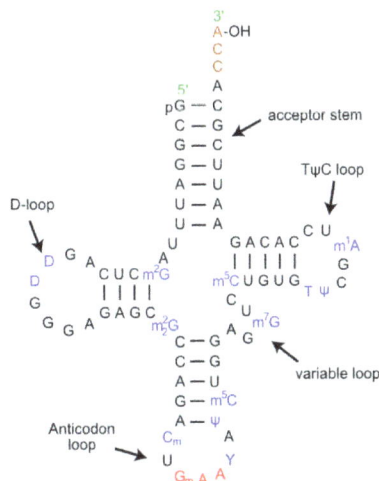

Secondary *cloverleaf structure* of tRNA^{Phe} from yeast.

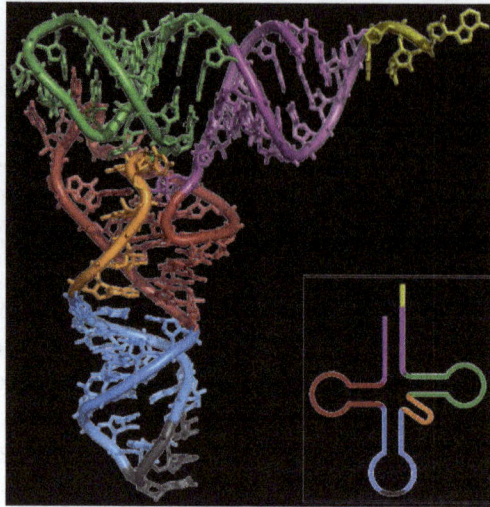

Tertiary structure of tRNA. *CCA tail* in yellow, *Acceptor stem* in purple, *Variable loop* in orange, *D arm* in red, *Anticodon arm* in blue with *Anticodon* in black, *T arm* in green.

The structure of tRNA can be decomposed into its primary structure, its secondary structure (usually visualized as the *cloverleaf structure*), and its tertiary structure (all tRNAs have a similar L-shaped 3D structure that allows them to fit into the P and A sites of the ribosome). The cloverleaf structure becomes the 3D L-shaped structure through coaxial stacking of the helices, which is a common RNA tertiary structure motif.

The lengths of each arm, as well as the loop 'diameter', in a tRNA molecule vary from species to species.

The tRNA structure consists of the following:

1. A 5'-terminalphosphate group.

2. The acceptor stem is a 7- to 9-base pair (bp) stem made by the base pairing of the 5'-terminal nucleotide with the 3'-terminal nucleotide (which contains the CCA 3'-terminal group used to attach the amino acid). The acceptor stem may contain non-Watson-Crick base pairs.

3. The CCA tail is a cytosine-cytosine-adenine sequence at the 3' end of the tRNA molecule. The amino acid loaded onto the tRNA by aminoacyl tRNA synthetases, to form aminoacyl-tRNA, is covalently bonded to the 3'-hydroxyl group on the CCA tail. This sequence is important for the recognition of tRNA by enzymes and critical in translation. In prokaryotes, the CCA sequence is transcribed in some tRNA sequences. In most prokaryotic tRNAs and eukaryotic tRNAs, the CCA sequence is added during processing and therefore does not appear in the tRNA gene.

4. The D arm is a 4- to 6-bp stem ending in a loop that often contains dihydrouridine.

5. The anticodon arm is a 5-bp stem whose loop contains the anticodon. The tRNA 5'-to-3' primary structure contains the anticodon but in reverse order, since 3'-to-5' directionality is required to read the mRNA from 5'-to-3'.

6. The T arm is a 4- to 5- bp stem containing the sequence TΨC where Ψ is pseudouridine, a modified uridine.

7. Bases that have been modified, especially by methylation (e.g. tRNA (guanine-N7-)-methyltransferase), occur in several positions throughout the tRNA. The first anticodon base, or wobble-position, is sometimes modified to inosine (derived from adenine), pseudouridine or lysidine (derived from cytosine).

Anticodon

An anticodon is a unit made up of three nucleotides that correspond to the three bases of the codon on the mRNA. Each tRNA contains a distinct anticodon triplet sequence that can base-pair to one or more codons for an amino acid. Some anticodons can pair with more than one codon due to a phenomenon known as wobble base pairing. Frequently, the first nucleotide of the anticodon is one not found on mRNA: inosine, which can hydrogen bond to more than one base in the corresponding codon position. In the genetic code, it is common for a single amino acid to be specified by all four third-position possibilities, or at least by both pyrimidines and purines; for example, the amino acid glycine is coded for by the codon sequences GGU, GGC, GGA, and GGG. Other modified nucleotides may also appear at the first anticodon position - sometimes known as the "wobble position" - resulting in subtle changes to the genetic code, as for example in mitochondria.

To provide a one-to-one correspondence between tRNA molecules and codons that specify amino acids, 61 types of tRNA molecules would be required per cell. However, many cells contain fewer than 61 types of tRNAs because the wobble base is capable of binding to several, though not necessarily all, of the codons that specify a particular amino acid. A minimum of 31 tRNA are required to translate, unambiguously, all 61 sense codons of the standard genetic code.

Aminoacylation

Aminoacylation is the process of adding an aminoacyl group to a compound. It covalently links an amino acid to the CCA 3' end of a tRNA molecule.

Each tRNA is aminoacylated (or *charged*) with a specific amino acid by an aminoacyl tRNA synthetase. There is normally a single aminoacyl tRNA synthetase for each amino acid, despite the fact that there can be more than one tRNA, and more than one anticodon, for an amino acid. Recognition of the appropriate tRNA by the synthetases is not mediated solely by the anticodon, and the acceptor stem often plays a prominent role.

Reaction:

1. amino acid + ATP → aminoacyl-AMP + PPi

2. aminoacyl-AMP + tRNA → aminoacyl-tRNA + AMP

Certain organisms can have one or more aminoacyl tRNA synthetases missing. This leads to charging of the tRNA by a chemically related amino acid. An enzyme or enzymes modify the charged amino acid to the final one. For example, *Helicobacter pylori* has glutaminyl tRNA synthetase missing. Thus, glutamate tRNA synthetase charges tRNA-glutamine(tRNA-Gln) with glutamate. An amidotransferase then converts the acid side chain of the glutamate to the amide, forming the correctly charged gln-tRNA-Gln.

Binding to Ribosome

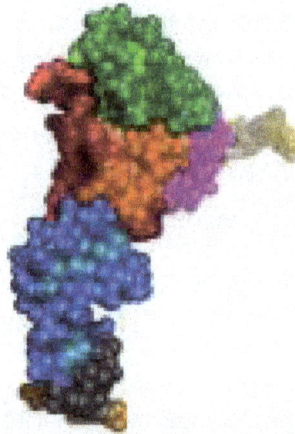

The range of conformations adopted by tRNA as it transits the A/T through P/E sites on the ribosome. The Protein Data Bank (PDB) codes for the structural models used as end points of the animation are given. Both tRNAs are modeled as phenylalanine-specific tRNA from *Escherichia coli*, with the A/T tRNA as a homology model of the deposited coordinates. Color coding as shown for tRNA tertiary structure. Adapted from.

The ribosome has three binding sites for tRNA molecules that span the space between the two ribosomal subunits: the A (aminoacyl),P (peptidyl), and E (exit) sites. In addition, the ribosome has two other sites for tRNA binding that are used during mRNA decoding or during the initiation of protein synthesis. These are the T site (named elongation factor Tu) and I site (initiation). By convention, the tRNA binding sites are denoted with the site on the small ribosomal subunit listed first and the site on the large ribosomal subunit listed second. For example, the A site is often written A/A, the P site, P/P, and the E site, E/E. The binding proteins like L27, L2, L14, L15, L16 at the A- and P- sites have been determined by affinity labeling by A.P. Czernilofsky et al. (*Proc. Natl. Acad. Sci, USA*, pp 230–234, 1974).

Once translation initiation is complete, the first aminoacyl tRNA is located in the P/P site, ready for the elongation cycle described below. During translation elongation, tRNA first binds to the ribosome as part of a complex with elongation factor Tu (EF-Tu) or its eukaryotic (eEF-1) or archaeal counterpart. This initial tRNA binding site is called the A/T site. In the A/T site, the A-site half resides in the small ribosomal subunit where the mRNA decoding site is located. The mRNA decoding site is where the mRNAcodon is read out during translation. The T-site half resides mainly on the large ribosomal subunit where EF-Tu or eEF-1 interacts with the ribosome. Once mRNA decoding is complete, the aminoacyl-tRNA is bound in the A/A site and is ready for the next peptide bond to be formed to its attached amino acid. The peptidyl-tRNA, which transfers the growing polypeptide to the aminoacyl-tRNA bound in the A/A site, is bound in the P/P site. Once the peptide bond is formed, the tRNA in the P/P site is deacylated, or has a free 3' end, and the tRNA in the A/A site carries the growing polypeptide chain. To allow for the next elongation cycle, the tRNAs then move through hybrid A/P and P/E binding sites, before completing the cycle and residing in the P/P and E/E sites. Once the A/A and P/P tRNAs have moved to the P/P and E/E sites, the mRNA has also moved over by one codon and the A/T site is vacant, ready for the next round of mRNA decoding. The tRNA bound in the E/E site then leaves the ribosome.

The P/I site is actually the first to bind to aminoacyl tRNA, which is delivered by an initiation factor called IF2 in bacteria. However, the existence of the P/I site in eukaryotic or archaeal ribosomes has not yet been confirmed. The P-site protein L27 has been determined by affinity labeling by E. Collatz and A.P. Czernilofsky (*FEBS Lett.*, Vol. 63, pp 283–286, 1976).

tRNA Genes

Organisms vary in the number of tRNA genes in their genome. The nematode worm *C. elegans*, a commonly used model organism in genetics studies, has 29,647 genes in its nuclear genome, of which 620 code for tRNA. The budding yeast *Saccharomyces cerevisiae* has 275 tRNA genes in its genome.

In the human genome, which, according to January 2013 estimates, has about 20,848 protein coding genes in total, there are 497 nuclear genes encoding cytoplasmic tRNA molecules, and 324 tRNA-derived pseudogenes—tRNA genes thought to be no longer functional (although pseudo tRNAs have been shown to be involved in antibiotic resistance in bacteria). Regions in nuclear chromosomes, very similar in sequence to mitochondrial tRNA genes, have also been identified (tRNA-lookalikes). These tRNA-lookalikes are also considered part of the nuclear mitochondrial DNA (genes transferred from the mitochondria to the nucleus).

As with all eukaryotes, there are 22 mitochondrial tRNA genes in humans. Mutations in some of these genes have been associated with severe diseases like the MELAS syndrome.

Cytoplasmic tRNA genes can be grouped into 49 families according to their anticodon features. These genes are found on all chromosomes, except 22 and Y chromosome. High clustering on 6p is observed (140 tRNA genes), as well on 1 chromosome.

The HGNC, in collaboration with the Genomic tRNA Database (GtRNAdb) and experts in the field, has approved unique names for human genes that encode tRNAs.

Evolution

Genomic tRNA content is a differentiating feature of genomes among biological domains of life: Archaea present the simplest situation in terms of genomic tRNA content with a uniform number of gene copies, Bacteria have an intermediate situation and Eukarya present the most complex situation. Eukarya present not only more tRNA gene content than the other two kingdoms but also a high variation in gene copy number among different isoacceptors, and this complexity seem to be due to duplications of tRNA genes and changes in anticodon specificity.

Evolution of the tRNA gene copy number across different species has been linked to the appearance of specific tRNA modification enzymes (uridine methyltransferases in Bacteria, and adenosine deaminases in Eukarya), which increase the decoding capacity of a given tRNA. As an example, tRNAAla encodes four different tRNA isoacceptors (AGC, UGC, GGC and CGC). In Eukarya, AGC isoacceptors are extremely enriched in gene copy number in comparison to the rest of isoacceptors, and this has been correlated with its A-to-I modification of its wobble base. This same trend has been shown for most amino acids of eukaryal species. Indeed, the effect of these two tRNA modifications is also seen in codon usage bias. Highly expressed genes seem to be enriched in codons that are exclusively using codons that will be decoded by these modified tRNAs, which suggests a possible role of these codons—and consequently of these tRNA modifications—in translation efficiency.

tRNA-derived Fragments

tRNA-derived fragments (or tRFs) are short molecules that emerge after cleavage of the mature tRNAs or the precursor transcript. Both cytoplasmic and mitochondrial tRNAs can produce fragments. There are at least four structural types of tRFs believed to originate from mature tRNAs, including the relatively long tRNA halves and short 5'-tRFs, 3'-tRFs and i-tRFs. The precursor tRNA can be cleaved to produce molecules from the 5' leader or 3' trail sequences. Cleavage enzymes include Angiogenin, Dicer, RNase Z and RNase P. Especially in the case of Angiogenin, the tRFs have a characteristically unusual cyclic phosphate at their 3' end and a hydroxyl group at the 5' end.

tRFs have multiple dependencies and roles. They exhibit significant changes between sexes, among races and disease status. Functionally, they can be loaded on Ago and act through RNAi pathways, participate in the formation of stress granules, displace mR-

NAs from RNA-binding proteins or inhibit translation. At the system or the organismal level, the four types of tRFs have a diverse spectrum of activities. Functionally, tRFs are associated with viral infection, cancer, cell proliferation and also with epigenetic transgenerational regulation of metabolism.

tRFs are not restricted to humans but have been shown to exist in multiple organisms.

Two online tools are available for those wishing to learn more about tRFs: the framework for the interactive exploration of mitochondrial and nuclear tRNA fragments (MINTbase) and the relational database of Transfer RNA related Fragments(tRFdb),. MINTbase also provides a naming scheme for the naming of tRFs called tRF-license plates that is genome independent.

tRNA Biogenesis

In eukaryotic cells, tRNAs are transcribed by RNA polymerase III as pre-tRNAs in the nucleus. RNA polymerase III recognizes two highly conserved downstream promoter sequences: the 5' intragenic control region (5'-ICR, D-control region, or A box), and the 3'-ICR (T-control region or B box) inside tRNA genes. The first promoter begins at +8 of mature tRNAs and the second promoter is located 30-60 nucleotides downstream of the first promoter. The transcription terminates after a stretch of four or more thymidines.

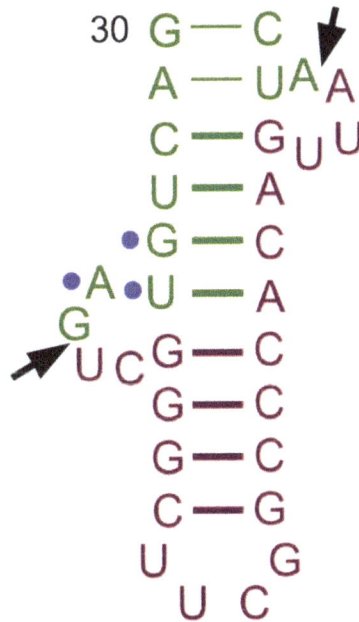

Bulge-helix-bulge motive of tRNA intron

Pre-tRNAs undergo extensive modifications inside the nucleus. Some pre-tRNAs contain introns that are spliced, or cut, to form the functional tRNA molecule; in bacteria these self-splice, whereas in eukaryotes and archaea they are removed by tRNA-splicing endonucleases. Eukaryotic pre-tRNA contains bulge-helix-bulge (BHB) structure motif that is important for recognition and precise splicing of tRNA intron by endonucleases. This motif position and structure are evolutionary conserved. However, some organisms, such as unicellular algae have a non-canonical position of BHB-motif as well as 5'- and 3'-ends of the spliced intron sequence. The 5' sequence is removed by RNase P, whereas the 3' end is removed by the tRNase Z enzyme. A notable exception is in the archaeon*Nanoarchaeum equitans,* which does not possess an RNase P enzyme and has a promoter placed such that transcription starts at the 5' end of the mature tRNA. The non-templated 3' CCA tail is added by a nucleotidyl transferase. Before tRNAs are exported into the cytoplasm by Los1/Xpo-t, tRNAs are aminoacylated. The order of the processing events is not conserved. For example, in yeast, the splicing is not carried out in the nucleus but at the cytoplasmic side of mitochondrial membranes.

History

The existence of tRNA was first hypothesized by Francis Crick, based on the assumption that there must exist an adapter molecule capable of mediating the translation of the RNA alphabet into the protein alphabet. Significant research on structure was conducted in the early 1960s by Alex Rich and Don Caspar, two researchers in Boston, the Jacques Fresco group in Princeton University and a United Kingdom group at King's College London. In 1965, Robert W. Holley of Cornell University reported the primary structure and suggested three secondary structures. tRNA was first crystallized in Madison, Wisconsin, by Robert M. Bock. The cloverleaf structure was ascertained by several other studies in the following years and was finally confirmed using X-ray crystallography studies in 1974. Two independent groups, Kim Sung-Houworking under Alexander Rich and a British group headed by Aaron Klug, published the same crystallography findings within a year.

References

- Danson, Michael; Eisenthal, Robert (2002). Enzyme assays: a practical approach. Oxford [Oxfordshire]: Oxford University Press. ISBN 0-19-963820-9.

- Gibson QH (1969). "Rapid mixing: Stopped flow". Methods in Enzymology. Methods in Enzymology. 16: 187–228. doi:10.1016/S0076-6879(69)16009-7. ISBN 978-0-12-181873-9.

- Duggleby RG (1995). "Analysis of enzyme progress curves by non-linear regression". Methods in Enzymology. Methods in Enzymology. 249: 61–90. doi:10.1016/0076-6879(95)49031-0. ISBN 978-0-12-182150-0. PMID 7791628.

- Fersht, Alan (1999). Structure and mechanism in protein science: a guide to enzyme catalysis and protein folding. San Francisco: W.H. Freeman. ISBN 0-7167-3268-8.

- Walsh, Ryan (2012). "Ch. 17. Alternative Perspectives of Enzyme Kinetic Modeling". In Ekinci, Deniz. Medicinal Chemistry and Drug Design(PDF). InTech. pp. 357–371. ISBN 978-953-51-0513-8.

- Nelson DL, Cox MM (2000). Lehninger Principles of Biochemistry (3rd ed.). New York: Worth Publishers. p. 206. ISBN 0-7167-6203-X.

- Jones, Daniel (2003) [1917], Peter Roach, James Hartmann and Jane Setter, eds., English Pronouncing Dictionary, Cambridge: Cambridge University Press, ISBN 3-12-539683-2

- Maton, Anthea; Jean Hopkins; Charles William McLaughlin; Susan Johnson; Maryanna Quon Warner; David LaHart; Jill D. Wright (1993). Human Biology and Health. Englewood Cliffs, New Jersey, USA: Prentice Hall. ISBN 0-13-981176-1.

- Steinberg, MH (2001). Disorders of Hemoglobin: Genetics, Pathophysiology, and Clinical Management. Cambridge University Press. p. 95. ISBN 0-521-63266-8.

- Ahrens; Kimberley, Basham (1993). Essentials of Oxygenation: Implication for Clinical Practice. Jones & Bartlett Learning. p. 194. ISBN 0867203323.

- Hall, John E. (2010). Guyton and Hall textbook of medical physiology (12th ed.). Philadelphia, Pa.: Saunders/Elsevier. p. 502. ISBN 978-1416045748.

- Guyton, Arthur C.; John E. Hall (2006). Textbook of Medical Physiology (11 ed.). Philadelphia: Elsevier Saunders. p. 511. ISBN 0-7216-0240-1.

- William, C. Wilson; Grande, Christopher M.; Hoyt, David B. (2007). "Pathophysiology of acute respiratory failure". Trauma, Volume II: Critical Care. Taylor & Francis. p. 430. ISBN 978-1-4200-1684-0.

- Handin, Robert I.; Lux, Samuel E. and StosselBlood, Thomas P. (2003). Blood: Principles & Practice of Hematology. Lippincott Williams & Wilkins, ISBN 0781719933

Quantum Biology: An Integrated Study

5

Quantum biology is the application of quantum mechanics to objects and problems. The processes involved are chemical reactions, formation of excited electronic states and transfer of excitation energy. The topics explained in the following chapter are visual phototransduction, magnetoreception, orchestrated objective reduction and biomagnetism.

Quantum Biology

Quantum biology refers to applications of quantum mechanics and theoretical chemistry to biological objects and problems. Many biological processes involve the conversion of energy into forms that are usable for chemical transformations and are quantum mechanical in nature. Such processes involve chemical reactions, light absorption, formation of excited electronic states, transfer of excitation energy, and the transfer of electrons and protons (hydrogen ions) in chemical processes such as photosynthesis and cellular respiration. Quantum biology may use computations to model biological interactions in light of quantum mechanical effects. Quantum biology is concerned with the influence of non-trivial quantum phenomena, as opposed to the so-called trivial quantum phenomena present in all biology by reduction to fundamental physics.

History

Early pioneers of quantum physics saw applications of quantum mechanics in biological problems. Erwin Schrödinger published *What is Life?* in 1944 discussing applications of quantum mechanics in biology. Schrödinger introduced the idea of an "aperiodic crystal" that contained genetic information in its configuration of covalent chemical bonds. He further suggested that mutations are introduced by "quantum leaps." Other pioneers Niels Bohr, Pascual Jordan, and Max Delbruck argued that the quantum idea of complementarity was fundamental to the life sciences. In 1963 Per-Olov Löwdin published proton tunneling as another mechanism for DNA mutation. In his paper, he stated that there is a new field of study called "quantum biology."

Applications

Photosynthesis

Organisms that undergo photosynthesis initially absorb light energy through the process of electron excitation in an antenna. This antenna varies between organisms. Bacteria can use ring like structures as antennas, whereas plants and other organisms use chlorophyll pigments to absorb photons. This electron excitation creates a separation of charge in a reaction site that is later converted into chemical energy for the cell to use. However, this electron excitation must be transferred in an efficient and timely manner, before that energy is lost in fluorescence.

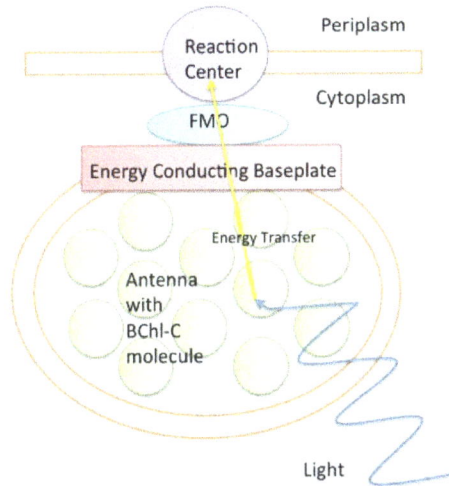

Diagram of FMO complex. Light excites electrons in an antenna. The excitation then transfers through various proteins in the FMO complex to the reaction center to further photosynthesis.

Various structures are responsible for transferring energy from the antennas to a reaction site. One of the most well studied is the FMO complex in green sulfur bacteria. FT electron spectroscopy studies show an efficiency of above 99% between the absorption of electrons and transfer to the reaction site with short lived intermediates. This high efficiency cannot be explained by classical mechanics such as a diffusion model.

Studies published in 2010 claimed to identify electronic quantum coherence and entanglement between the excited states of different pigments in the light-harvesting stage of photosynthesis. However, critical follow-up studies question the interpretation of these results and assign the reported signatures of electronic quantum coherence to nuclear dynamics in the chromophores. There are a number of proposals as to how quantum coherence transfers the absorbed energy to the reaction site. According to one proposal, if each site within the complex feels its own environmental noise, then because of both quantum coherence and thermal environment, the electron will not remain in any local minimum but proceed to the reaction site. Another proposal is that the rate of quantum coherence combined with electron tunneling creates an energy sink that moves the electron to the reaction site quickly. Recent work indicates that symmetries present in the geometric arrangement of the complex may favor efficient energy transfer to the reaction center, in a way that resembles perfect state transfer in quantum networks. However, control experiments cast doubts on the interpretation that quantum effects last any longer than one hundred femtoseconds.

Vision

Vision relies on quantized energy in order to convert light signals to an action potential in a process called phototransduction. In phototransduction, a photon interacts with a chromophore in a light receptor. The chromophore absorbs the photon and undergoes photoisomerization. This change in structure induces a change in the structure of the photo receptor and resulting signal transduction pathways lead to a visual signal. However, the photoisomerization reaction occurs at a rapid rate, <200 fs, with high yield. Models suggest the use of quantum effects in shaping the ground state and excited state potentials in order to achieve this efficiency.

Enzymatic Activity (Quantum Biochemistry)

Enzymes may use quantum tunneling to transfer electrons long distances. Tunneling refers to the ability of a small mass particle to travel through energy barriers. Studies show that long distance electron transfers between redox centers through quantum tunneling plays important roles in enzymatic activity of photosynthesis and cellular respiration. For example, studies show that long range electron tunneling on the order of 15–30 Å plays a role in redox reactions in enzymes of cellular respiration. Even though there are such large separations between redox sites within enzymes, electrons successfully transfer in a temperature independent and distance dependent manner. This suggests the ability of electrons to tunnel in physiological conditions. Further research is needed to determine whether this specific tunneling is also coherent.

Magnetoreception

Magnetoreception refers to the ability of animals to navigate using the inclination of the magnetic field of the earth. A possible explanation for magnetoreception is the radical pair mechanism. The radical-pair mechanism is well-established in spin chemistry, and was speculated to apply to magnetoreception in 1978 by Schulten et al.. In 2000, cryptochrome was proposed as the "magnetic molecule," so to speak, that could harbor magnetically sensitive radical-pairs. Cryptochrome, a flavoprotein found in the eye of at least the European robin, is the only protein known to form photoinduced radical-pairs in animals. The function of cryptochrome is diverse across species, however, the photoinduction of radical-pairs occurs by exposure to blue light, which excites an electron in a chromophore.

Furthermore, the excited electron is quantum entangled to another electron so it is simultaneously in both the singlet and triplet states. However, in principle, entanglement is not necessary for a radical-pair mechanism of magnetoreception, so evidence of a radical-pair mechanism of magnetoreception is not necessarily synonymous with evidence for quantum biology.

Nevertheless, in the lab, the direction of weak magnetic fields can affect radical-pair's reactivity, and therefore can "catalyze" the formation of chemical products. Whether this mechanism applies to magnetoreception and/or quantum biology, that is, whether earth's magnetic field "catalyzes" the formation of *bio*chemical products by the aid of entangled or non-entangled radical-pairs, is doubly undetermined. As to the former, researchers found evidence for the radical-pair mechanism of magnetoreception when European robins, cockroaches, and garden warblers, could no longer navigate when exposed to a radio frequency oscillating magnetic fields, which specially disturbs radical-pair chemistry. To empirically suggest the involvement of entanglement, an experiment would need to be devised that could disturb entangled radical-pairs without disturbing other radical-pairs, or vice versa, which would first need to be demonstrated in a laboratory setting before being applied to magnetoreception.

Other Biological Applications

Other examples of quantum phenomena in biological systems include olfaction, the conversion of chemical energy into motion, DNA mutation and brownian motors in many cellular processes.

Visual Phototransduction

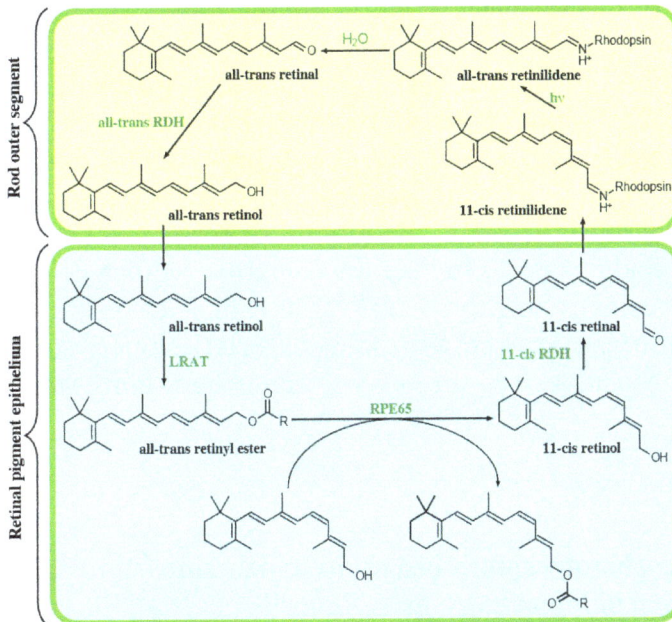

The Visual Cycle.   hv = Incident photon

Visual phototransduction is the sensory transduction of the visual system. It is a process by which light is converted into electrical signals in the rod cells, cone cells and photosensitive ganglion cells of the retina of the eye.This cycle was elucidated by George Wald(1906-1997) for which he received the Nobel Prize in 1967.It is so called "Wald's Visual Cycle" after him.

The visual cycle is the biological conversion of a photon into an electrical signal in the retina. This process occurs via G-protein coupled receptors called opsins which contain the chromophore11-cis retinal. 11-cis retinal is covalently linked to the opsin receptor via Schiff base forming retinylidene protein. When struck by a photon, 11-cis retinal undergoes photoisomerization to all-trans retinal which changes the conformation of the opsinGPCR leading to signal transduction cascades which causes closure of cyclic GMP-gated cation channel, and hyperpolarization of the photoreceptor cell.

Following isomerization and release from the opsinprotein, all-trans retinal is reduced to all-trans retinol and travels back to the retinal pigment epithelium to be "recharged". It is first esterified by lecithin retinol acyltransferase (LRAT) and then converted to 11-cis retinol by the isomerohydrolase RPE65. The isomerase activity of RPE65 has been shown; it is still uncertain whether it also acts as hydrolase. Finally, it is oxidized to 11-cis retinal before traveling back to the rod outer segment where it is again conjugated to an opsin to form new, functional visual pigment (rhodopsin).

Photoreceptors

The photoreceptor cells involved in vision are the rods and cones. These cells contain a chromophore (11-cis retinal, the aldehyde of Vitamin A1 and light-absorbing portion) bound to cell membrane protein, opsin. Rods deal with low light level and do not mediate color vision. Cones, on the other hand, can code the color of an image through comparison of the outputs of the three different types of cones. Each cone type responds best to certain wavelengths, or colors, of light because each type has a slightly different opsin. The three types of cones are L-cones, M-cones and S-cones that respond optimally to long wavelengths (reddish color), medium wavelengths (greenish color), and short wavelengths (bluish color) respectively. Humans have a trichromatic visual system consisting of three unique systems, rods, mid and long-wavelength sensitive (red and green) cones and short wavelength sensitive (blue) cones.

Process

To understand the photoreceptor's behaviour to light intensities, it is necessary to understand the roles of different currents.

There is an ongoing outward potassium current through nongated K^+-selective channels. This outward current tends to hyperpolarize the photoreceptor at around -70 mV (the equilibrium potential for K^+).

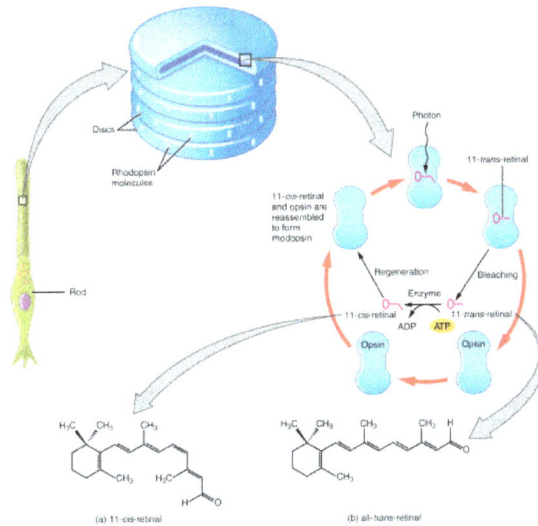

The absorption of light leads to an isomeric change in the retinal molecule.

There is also an inward sodium current carried by cGMP-gated sodium channels. This so-called 'dark current' depolarizes the cell to around -40 mV. Note that this is significantly more depolarized than most other neurons.

A high density of Na$^+$-K$^+$ pumps enables the photoreceptor to maintain a steady intracellular concentration of Na$^+$ and K$^+$.

In the Dark

Photoreceptor cells are unusual cells in that they depolarize in response to absence of stimuli or scotopic conditions (darkness). In photopic conditions (light), photoreceptors hyperpolarize to a potential of -60mV. It is this 'switching off' that activates the next cell and sends an excitatory signal down the neural pathway.

In the dark, cGMP levels are high and keep cGMP-gated sodium channels open allowing a steady inward current, called the dark current. This dark current keeps the cell depolarized at about -40 mV.

The depolarization of the cell membrane in scotopic conditions opens voltage-gated calcium channels. An increased intracellular concentration of Ca^{2+} causes vesicles containing neurotransmitters, to merge with the cell membrane, therefore releasing the neurotransmitter into the synaptic cleft, an area between the end of one cell and the beginning of another neuron. The neurotransmitter released is glutamate, a neurotransmitter whose receptors are often excitatory.

In the cone pathway glutamate:

- Hyperpolarizes on-center bipolar cells. Glutamate that is released from the photoreceptors in the dark binds to metabotropic glutamate receptors (mGluR6), which, through a G-protein coupling mechanism, causes non-specific cation channels in the cells to close, thus hyperpolarizing the bipolar cell.

- Depolarizes off-center bipolar cells. Binding of glutamate to ionotropic glutamate receptors results in an inward cation current that depolarizes the bipolar cell.

In the Light

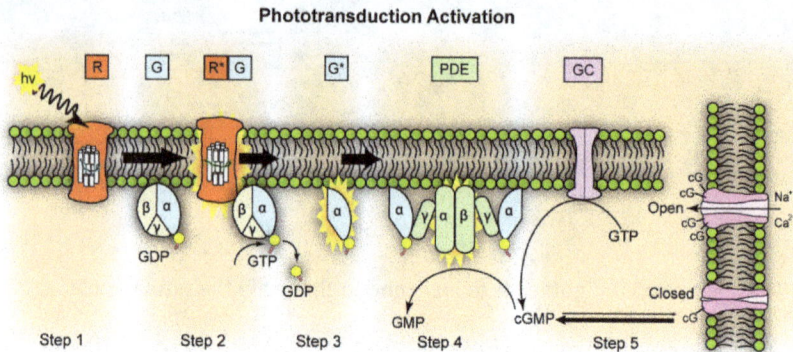

Phototransduction Activation

Representation of molecular steps in photoactivation (modified from Leskov et al., 2000). Depicted is an outer membrane disk in a rod. Step 1: Incident photon (hv) is absorbed and activates a rhodopsin by conformational change in the disk membrane to R*. Step 2: Next, R* makes repeated contacts with transducin molecules, catalyzing its activation to G* by the release of bound GDP in exchange for cytoplasmic GTP, which expels its β and γ subunits. Step 3: G* binds inhibitory γ subunits of the phosphodiesterase (PDE) activating its α and β subunits. Step 4: Activated PDE hydrolyzes cGMP. Step 5: Guanylyl cyclase (GC) synthesizes cGMP, the second messenger in the phototransduction cascade. Reduced levels of cytosolic cGMP cause cyclic nucleotide gated channels to close preventing further influx of Na+ and Ca2+.

1. A light photon interacts with the retinal in a photoreceptor cell. The retinal undergoes isomerisation, changing from the 11-*cis* to all-*trans* configuration

2. Retinal no longer fits into the opsin binding site.

3. Opsin therefore undergoes a conformational change to metarhodopsin II.

4. Metarhodopsin II is unstable and splits, yielding opsin and all-*trans* retinal.

5. The opsin activates the regulatory protein transducin. This causes transducin to dissociate from its bound GDP, and bind GTP, then the alpha subunit of transducin dissociates from the beta and gamma subunits, with the GTP still bound to the alpha subunit.

6. The alpha subunit-GTP complex activates phosphodiesterase or PDE.

7. PDE breaks down cGMP to 5'-GMP. This lowers the concentration of cGMP and therefore the sodium channels close.

8. Closure of the sodium channels causes hyperpolarization of the cell due to the ongoing efflux of potassium ions.

9. Hyperpolarization of the cell causes voltage-gated calcium channels to close.

10. As the calcium level in the photoreceptor cell drops, the amount of the neu-rotransmitter glutamate that is released by the cell also drops. This is because calcium is required for the glutamate-containing vesicles to fuse with cell membrane and release their contents.

11. A decrease in the amount of glutamate released by the photoreceptors causes depolarization of On center bipolar cells (rod and cone On bipolar cells) and hyperpolarization of cone off-center bipolar cells.

Deactivation of the Phototransduction Cascade

GTPase Accelerating Protein (GAP) interacts with the alpha subunit of transducin, and causes it to hydrolyse its bound GTP to GDP, and thus halts the action of phosphodiesterase, stopping the transformation of cGMP to GMP.

Guanylate Cyclase Activating Protein (GCAP) is a calcium binding protein, and as the calcium levels in the cell have decreased, GCAP dissociates from its bound calcium ions, and interacts with Guanylate Cyclase, activating it. Guanylate Cyclase then proceeds to transform GTP to cGMP, replenishing the cell's cGMP levels and thus reopening the sodium channels that were closed during phototransduction.

Finally, Metarhodopsin II is deactivated. Recoverin, another calcium binding protein, is normally bound to Rhodopsin Kinase when calcium is present. When the calcium levels fall during phototransduction, the calcium dissociates from recoverin, and rhodopsin kinase is released, when it proceeds to phosphorylate metarhodopsin II, which decreases its affinity for transducin. Finally, arrestin, another protein, binds the phosphorylated metarhodopsin II, completely deactivating it. Thus, finally, phototransduction is deactivated, and the dark current and glutamate release is restored. It is this pathway, where Metarhodopsin II is phosphorylated and bound to arrestin and thus deactivated, which is thought to be responsible for the S2 component of dark adaptation. The S2 component represents a linear section of the dark adaptation function present at the beginning of dark adaptation for all bleaching intensities.

All-*trans* retinal is transported to the pigment epithelial cells to be reduced to all-*trans* retinol, the precursor to 11-*cis* retinal. This is then transported back to the rods. All-*trans* retinal cannot be synthesised by humans and must be supplied by vitamin A in the diet. Deficiency of all-*trans* retinal can lead to night blindness. This is part of the bleach and recycle process of retinoids in the photoreceptors and retinal pigment epithelium.

Phototransduction in Invertebrates

Phototransduction process in invertebrates like the fruit fly is different from the vertebrates. Here, light induces the conformational change into Rhodopsin and converts it into meta-rhodopsin. This helps in dissociation of G -protein complex. Alpha sub-unit of this complex activates the PLC enzyme (PLC-beta) which hydrolyze the PIP2 into DAG. This hydrolysis leads to opening of TRP channels and influx of calcium.

Magnetoreception

The homing pigeon can return to its home using its ability to sense the
Earth's magnetic field and other cues to orient itself.

Magnetoreception (also magnetoception) is a sense which allows an organism to detect a magnetic field to perceive direction, altitude or location. This sensory modality is used by a range of animals for orientation and navigation, and as a method for animals to develop regional maps. For the purpose of navigation, magnetoreception deals with the detection of the Earth's magnetic field.

Magnetoreception is present in bacteria, arthropods, molluscs and members of all major taxonomic groups of vertebrates. Humans are not thought to have a magnetic sense, but there is a protein (a cryptochrome) in the eye which could serve this function.

Proposed Mechanisms

An unequivocal demonstration of the use of magnetic fields for orientation within an organism has been in a class of bacteria known as magnetotactic bacteria. These bacteria demonstrate a behavioural phenomenon known as magnetotaxis, in which the bacterium orients itself and migrates in the direction along the Earth's magnetic field lines. The bacteria contain magnetosomes, which are nanometer-sized particles of magnetite or iron sulfide enclosed within the bacterial cells. The magnetosomes are surrounded by a membrane composed of phospholipids and fatty acids and contain at least 20 different proteins They form in chains where the moments of each magnetosome align in parallel, causing each bacterium cell to essentially act as a magnetic dipole, giving the bacteria their permanent-magnet characteristics.

For animals the mechanism for magnetoreception is unknown, but there exist two main hypotheses to explain the phenomenon. According to one model, cryptochrome, when exposed to blue light, becomes activated to form a pair of radicals (molecules with a single unpaired electron) where the spins of the two unpaired electrons are correlated. The surrounding magnetic field affects the dynamics of this correlation (parallel or anti-parallel), and this in turn affects the length of time cryptochrome stays in its activated state. Activation of cryptochrome may affect the light-sensitivity of retinal neurons, with the overall result that the bird can see the color phase shift caused by the magnetic field. The Earth's magnetic field is only 0.5 Gauss and so it is difficult to conceive of a mechanism, other than phase shift, by which such a field could lead to any chemical changes other than those affecting the weak magnetic fields between radical pairs. Cryptochromes are therefore thought to be essential for the light-dependent ability of the fruit fly *Drosophila melanogaster* to sense magnetic fields.

The second proposed model for magnetoreception relies on Fe_3O_4, also referred to as iron (II, III) oxide or magnetite, a natural oxide with strong magnetism. Iron (II, III) oxide remains permanently magnetized when its length is larger than 50 nm and becomes magnetized when exposed to a magnetic field if its length is less than 50 nm. In both of these situations the Earth's magnetic field leads to a transducible signal via a physical effect on this magnetically sensitive oxide.

Another less general type of magnetic sensing mechanism in animals that has been thoroughly described is the inductive sensing methods used by sharks, stingrays and chimaeras (cartilaginous fish). These species possess a unique electroreceptive organ known as *ampullae of Lorenzini* which can detect a slight variation in electric potential. These organs are made up of mucus-filled canals that connect from the skin's pores to small sacs within the animal's flesh that are also filled with mucus. The ampullae of Lorenzini are capable of detecting DC currents and have been proposed to be used in the sensing of the weak electric fields of prey and predators. These organs could also possibly sense magnetic fields, by means of Faraday's law: as a conductor moves through a magnetic field an electric potential is generated. In this case the conductor is the animal moving through a magnetic field, and the potential induced depends on the time varying rate of flux through the conductor according to

These organs detect very small fluctuations in the potential difference between the pore and the base of the electroreceptor sack. An increase in potential results in a decrease in the rate of nerve activity, and a decrease in potential results in an increase in the rate of nerve activity. This is analogous to the behavior of a current carrying conductor; with a fixed channel resistance, an increase in potential would decrease the amount of current detected, and vice versa. These receptors are located along the mouth and nose of sharks and stingrays.

In Invertebrates

The nematode *Caenorhabditis elegans* was recently shown to orient to the magnetic field of the earth using the first described set of magnetosensory neurons. Worms appear to use the magnetic field to orient during vertical soil migrations that change in sign depending on their satiation state (with hungry worms burrowing down, and satiated worms burrowing up).

The mollusc *Tochuina tetraquetra* (formerly *Tritonia diomedea* or *Tritonia gigantea*) has been studied for clues as to the neural mechanism behind magnetoreception in a species. Some of the earliest work with *Tochuina* showed that prior to a full moon *Tochuina* would orient their bodies between magnetic north and east. A Y-maze was established with a right turn equal to geomagnetic south and a left turn equal to geomagnetic east. Within this geomagnetic field 80% of *Tochuina* made a turn to the left or magnetic east. However, when a reversed magnetic field was applied that rotated magnetic north 180° there was no significant preference for either turn, which now corresponded with magnetic north and magnetic west. These results, though interesting, do not conclusively establish that *Tochuina* uses magnetic fields in magnetoreception. These experiments do not include a control for the activation of the Rubens' coil in the reversed magnetic field experiments. Therefore, it is possible that heat or noise generated by the coil was responsible for the loss of choice preference. Further work with *Tochuina* was unable to identify any neurons that showed rapid changes in firing as a result of magnetic fields. However, pedal 5 neurons, two bisymmetric neurons located within the *Tochuina* pedal ganglion, exhibited gradual changes in firing over time following 30 minutes of magnetic stimulation provided by a Rubens' coil. Further studies showed that pedal 7 neurons in the pedal ganglion were inhibited when exposed to magnetic fields over the course of 30 minutes. The function of both pedal 5 neurons and pedal 7 neurons is currently unknown.

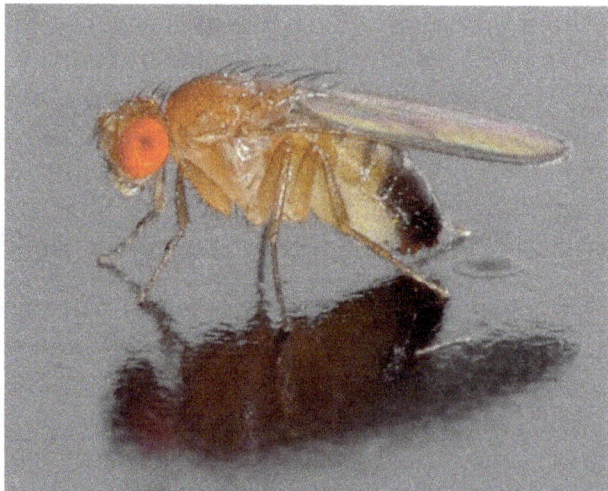

Fruit fly *Drosophila melanogaster* needs cryptochrome to respond to magnetic fields.

Drosophila melanogaster is another invertebrate which may be able to orient to magnetic fields. Experimental techniques such as gene knockouts have allowed a closer examination of possible magnetoreception in these fruit flies. Various *Drosophila* strains have been trained to respond to magnetic fields. In a choice test flies were loaded into an apparatus with two arms that were surrounded by electric coils. Current was run through each of the coils, but only one was configured to produce a 5-Gauss magnetic field at a time. The flies in this T-maze were tested on their native ability to recognize the presence of the magnetic field in an arm and on their response following training where the magnetic field was paired with a sucrose reward. Many of the strains of flies showed a learned preference for the magnetic field following training. However, when the only cryptochrome found in *Drosophila*, type 1 Cry, is altered, either through a missense mutation or replacement of the Cry gene, the flies exhibit a loss of magnetosensitivity. Furthermore, when light is filtered to only allow wavelengths greater than 420 nm through, *Drosophila* loses its trained response to magnetic fields. This response to filtered light is likely linked to the action spectrum of fly-cryptochrome which has a range from 350 nm – 400 nm and plateaus from 430-450 nm. Although researchers had believed that a tryptophan triad in cryptochrome was responsible for the free radicals on which magnetic fields could act, recent work with *Drosophila* has shown that tryptophan might not be behind cryptochrome dependent magnetoreception. Alteration of the tryptophan protein does not result in the loss of magnetosensitivity of a fly expressing either type 1 Cry or the cryptochrome found in vertebrates, type 2 Cry. Therefore, it remains unclear exactly how cryptochrome mediates magnetoreception. These experiments used a 5 gauss magnetic field, 10 times the strength of the Earth's magnetic field). *Drosophila* has not been shown to be able to respond to the Earth's weaker magnetic field.

Magnetoreception is well documented in honey bees, ants and termites. In ants and bees, this is used to orient and navigate in areas around their nests and within their migratory paths. For example, through the use of magnetoreception, the Brazilian stingless bee *Schwarziana quadripunctata* is able to distinguish differences in altitude, location, and directionality using the thousands of hair-like particles on its antennae.

In Homing Pigeons

Homing pigeons can use magnetic fields as part of their complex navigation system.William Keeton showed that time-shifted homing pigeons are unable to orient themselves correctly on a clear, sunny day which is attributed to time-shifted pigeons being unable to compensate accurately for the movement of the sun during the day. Conversely, time-shifted pigeons released on overcast days navigate correctly. This led to the hypothesis that under particular conditions, homing pigeons rely on magnetic fields to orient themselves. Further experiments with magnets attached to the backs of homing pigeons demonstrated that disruption of the bird's ability to sense the Earth's magnetic field leads to a loss of proper orientation behavior under overcast conditions. There have been two mechanisms implicated in homing pigeon magnetoreception : the visually mediated free-radical pair mechanism and a magnetite based directional compass or inclination compass. More recent behavioral tests have shown that pigeons are able to detect magnetic anomalies of 186 microtesla (1.86 Gauss).

In a choice test birds were trained to jump onto a platform on one end of a tunnel if there was no magnetic field present and to jump onto a platform on the other end of the tunnel if a magnetic field was present. In this test, birds were rewarded with a food prize and punished with a time penalty. Homing pigeons were able to make the correct choice 55%-65% of the time which is higher than what would be expected if the pigeons were simply guessing. The ability of pigeons to detect a magnetic field is impaired by application of lidocaine, an anesthetic, to the olfactory mucosa. Furthermore, sectioning the trigeminal nerve leads to an inability to detect a magnetic field, while sectioning of the olfactory nerve has no effect on the magnetic sense of homing pigeons. These results suggest that magnetite located in the beak of pigeons may be responsible for magnetoreception via trigeminal mediation. However, it has not been shown that the magnetite located in the beak of pigeons is capable of responding to a magnetic field with the Earth's strength. Therefore, the receptor responsible for magnetosensitivity in homing pigeons remains uncertain.

Aside from the sensory receptor for magnetic reception in homing pigeons there has been work on neural regions that are possibly involved in the processing of magnetic information within the brain. Areas of the brain that have shown increases in activity in response to magnetic fields with a strength of 50 or 150 microtesla are the posterior vestibular nuclei, dorsal thalamus, hippocampus, and visual hyperpallium.

Because pigeons provided some of the first evidence for the use of magnetoreception in navigation, they have been an organism of focus in magnetoreception studies. The precise mechanism used by pigeons has not been established and so it is unclear yet whether pigeons rely solely on a cryptochrome-mediated receptor or on beak magnetite.

In Domestic Hens

Domestic hens have iron mineral deposits in the dendrites in the upper beak and are capable of magnetoreception. Because hens use directional information from the magnetic field of the earth to orient in relatively small areas, this raises the possibility that beak-trimming (removal of part of the beak, to reduce injurious pecking, frequently performed on egg-laying hens) impairs the ability of hens to orient in extensive systems, or to move in and out of buildings in free-range systems.

In Mammals

Several mammals, including the big brown bat (*Eptesicus fuscus*) can use magnetic fields for orientation.

Work with mice, mole-rats and bats has shown that some mammals are capable of magnetoreception. When woodmice are removed from their home area and deprived of visual and olfactory cues, they orient themselves towards their homes until an inverted magnetic field is applied to their cage. However, when the same mice are allowed access to visual cues, they are able to orient themselves towards home despite the presence of inverted magnetic fields. This indicates that when woodmice are displaced, they use magnetic fields to orient themselves if there are no other cues available. However, early studies of these subjects were criticized because of the difficulty of completely removing sensory cues, and in some because the tests were performed out of the natural environment. In others, the results of these experiments do not conclusively show a response to magnetic fields when deprived of other cues, because the magnetic field was artificially changed before the test rather than during it.

Later research, accounting for those factors, has led to a conclusion that the Zambian mole-rat, a subterranean mammal, uses magnetic fields as a polarity compass to aid in the orientation of their nests. In contrast to work with woodmice, Zambian mole-rats do not exhibit different orientation behavior when a visual cue such as the sun is present, a result that has been suggested is due to their subterranean lifestyle. Further investigation of mole-rat magnetoreception lead to the finding that exposure to magnetic fields leads to an increase in neural activity within the superior colliculus as measured by immediate early gene expression. The activity level of neurons within two levels of the superior colliculus, the outer sublayer of the intermediate gray layer and the deep gray layer, were elevated in a non-specific manner when exposed to various magnetic fields. However, within the inner sublayer of the intermediate gray layer (InGi) there were two or three clusters of responsive cells. The more time the mole rats were exposed to a magnetic field the greater the immediate early gene expression within the InGi. However, if Zambian mole-rats were placed in a field with a shielded magnetic field only a few scattered cells were active. Therefore, it has been proposed that in mammals, the superior colliculus is an important neural structure in the processing of magnetic information.

Bats may also use magnetic fields to orient themselves. While it is known that bats use echolocation to navigate over short distances, it is unclear how they navigate over longer distances. When *Eptesicus fuscus* are taken from their home roosts and exposed to magnetic fields 90 degrees clockwise or counterclockwise of magnetic north, they are disoriented and set off for their homes in the wrong direction. Therefore, it seems that *Eptesicus fuscus* is capable of magnetoreception. However, the exact use of magnetic fields by *Eptesicus fuscus* is unclear as the magnetic field could be used either as a map, a compass, or a compass calibrator. Recent research with another bat species, *Myotis myotis*, supports the hypothesis that bats use magnetic fields as a compass calibrator and their primary compass is the sun.

Red foxes (*Vulpes vulpes*) may use magnetoreception when predating small rodents. When foxes perform their high-jumps onto small prey like mice and voles, they tend to jump in a north-eastern compass direction. In addition, successful attacks are "tightly clustered" to the north. One study has found that when domestic dogs are off the leash and the Earth's magnetic field is calm, they prefer to urinate and defecate with their bodies aligned on a north-south axis.

There is also evidence for magnetoreception in large mammals. Resting and grazing cattle as well as roe deer (*Capreolus capreolus*) and red deer (*Cervus elaphus*) tend to align their body axes in the geomagnetic north-south direction. Because wind, sunshine, and slope could be excluded as common ubiquitous factors in this study, alignment toward the vector of the magnetic field provided the most likely explanation for the observed behaviour. However, because of the descriptive nature of this study, alternative explanations (e.g., the sun compass) could not be excluded. In a follow-up study, researchers analyzed body orientations of ruminants in localities where the geomagnetic field is disturbed by high-voltage power lines to determine how local variation in magnetic fields may affect orientation behaviour. This was done by using satellite and aerial images of herds of cattle and field observations of grazing roe deer. Body orientation of both species was random on pastures under or near power lines. Moreover, cattle exposed to various magnetic fields directly beneath or in the vicinity of power lines trending in various magnetic directions exhibited distinct patterns of alignment. The disturbing effect of the power lines on body alignment diminished with the distance from the conductors. In 2011 a group of Czech researchers, however, reported their failed attempt to replicate the finding using different Google Earth images.

Humans "are not believed to have a magnetic sense", but humans do have a cryptochrome (a flavoprotein, CRY2) in the retina which has a light-dependent magnetosensitivity. CRY2 "has the molecular capability to function as a light-sensitive magnetosensor", so the area was thought (2011) to be ripe for further study.

Issues

The largest issue affecting verification of an animal magnetic sense is that despite more than 40 years of work on magnetoreception there has yet to be an identification of a sensory receptor. Given that the entire receptor system could likely fit in a one-millimeter cube and have a magnetic content of less than one ppm, it is difficult to discern the parts of the brain where this information is processed. In various organisms a cryptochrome mediated receptor has been implicated in magnetoreception. At the same time a magnetite system has been found to be relevant to magnetosensation in birds. Furthermore, it is possible that both of these mechanisms play a role in magnetic field detection in animals. This dual mechanism theory in birds raises the questions: If such a mechanism is actually responsible for magnetoreception, to what degree is each method responsible for stimulus transduction, and how do they lead to a tranducible signal given a magnetic field with the Earth's strength?

The precise purpose of magnetoreception in animal navigation is unclear. Some animals appear to use their magnetic sense as a map, compass, or compass calibrator. The compass method allows animals not only to find north, but also to maintain a constant heading in a particular direction. Although the ability to sense direction is important in migratory navigation, many animals also have the ability to sense small fluctuations in earth's magnetic field to compute coordinate maps with a resolution of a few kilometers or better. For example, birds such as the homing pigeon are believed to use the magnetite in their beaks to detect magnetic signposts and thus, the magnetic sense they gain from this pathway is a possible map. Yet, it has also been suggested that homing pigeons and other birds use the visually mediated cryptochrome receptor as a compass.

The purpose of magnetoreception in birds and other animals may be varied, but has proved difficult to study. Numerous studies use magnetic fields larger than the Earth's field. Studies such as of *Tritonia* have used electrophysiological recordings from only one or two neurons, and many others have been solely correlational.

Orchestrated Objective Reduction

Orchestrated objective reduction (Orch-OR) is a hypothesis that consciousness in the brain originates from processes *inside*neurons, rather than from connections *between* neurons (the conventional view). The mechanism is held to be a quantum physics process called objective reduction that is orchestrated by molecular structures called microtubules. Objective reduction is proposed to be influenced by non-computable factors imbedded in spacetime geometry which thus may account for the Hard Problem of Consciousness. The hypothesis was put forward in the early 1990s by theoretical physicistRoger Penrose and anaesthesiologist and psychologistStuart Hameroff.

Overview

The hypothesis states that consciousness derives from deeper-level, finer-scale quantum activities inside cells, most prevalent in the neurons. It combines approaches from molecular biology, neuroscience, quantum physics, pharmacology, philosophy, quantum information theory and from a yet to come theory to "gravitize" quantum mechanics.

While mainstream theories assert that consciousness emerges as the complexity of the computations performed by cerebralneurons increases, Orch-OR posits that consciousness is based on non-computablequantum processing performed by qubits formed collectively on cellular microtubules, a process significantly amplified in the neurons. The qubits are based on oscillating dipoles forming superposed resonance rings in helical pathways throughout microtubule lattices. The oscillations are either electric, due to charge separation from London forces, or (most favorably) magnetic, due to electron spin—and possibly also due to nuclear spins (that can remain isolated for longer periods) and that occur in gigahertz, megahertz and kilohertz frequency ranges. Orchestration refers to the hypothetical process by which connective proteins, such as microtubule-associated proteins (MAPs), influence or orchestrate qubit state reduction by modifying the spacetime-separation of their superimposed states. The latter is based on Penrose's objective collapse theory for interpreting quantum mechanics, which postulates the existence of an objective threshold governing the collapse of quantum-states, related to the difference of the space-time curvature of these states in the universe's fine scale structure.

The basis of Orch-OR has been criticized from its inception by mathematicians, philosophers, and scientists, prompting the authors to revise and elaborate many of the theory's peripheral assumptions, retaining the core hypothesis. The criticism concentrated on three issues: Penrose's interpretation of Gödel's theorem; Penrose's abductive reasoning linking non-computability to quantum processes; the brain's unsuitability to host the quantum phenomena required by the theory, since it is considered too *"warm, wet and noisy"* to avoid decoherence. In other words, there is a missing link between physics and neuroscience and to date, it is too premature to claim that the Orch-OR hypothesis is right or wrong.

Penrose–Lucas Argument

The Penrose–Lucas argument states that, because humans are capable of knowing the truth of Gödel-unprovable statements, human thought is necessarily non-computable.

In 1931, mathematician and logician Kurt Gödelproved that any effectively generated theory capable of proving basic arithmetic cannot be both consistent and complete. Furthermore, he showed that any such theory also including a statement of its own consistency is inconsistent. A key element of the proof is the use of Gödel numbering to construct a "Gödel sentence" for the theory, which encodes a statement of its own incompleteness, e.g. "This theory can't assert the truth of this statement." This statement is either true but unprovable (incompleteness) *or* false and provable (inconsistency). An analogous statement has been used to show that humans are subject to the same limits as machines.

However, in his first book on consciousness, *The Emperor's New Mind* (1989), Penrose made Gödel's theorem the basis of what quickly became a controversial claim. He argued that while a formal proof system cannot prove its own consistency, Gödel-unprovable results are provable by human mathematicians. He takes this disparity to mean that human mathematicians are not describable as formal proof systems, and are therefore running a non-computable algorithm. Similar claims about the implications of Gödel's theorem were originally espoused by the philosopher John Lucas of Merton College, Oxford.

The inescapable conclusion seems to be: Mathematicians are not using a knowably sound calculation procedure in order to ascertain mathematical truth. We deduce that mathematical understanding – the means whereby mathematicians arrive at their conclusions with respect to mathematical truth – cannot be reduced to blind calculation!

— Roger Penrose

Objective Reduction

Motivation

If correct, the Penrose–Lucas argument creates a need to understand the physical basis of non-computable behaviour in the brain. Most physical laws are computable, and thus algorithmic. However, Penrose determined that wave function collapse was a prime candidate for a non-computable process.

In quantum mechanics, particles are treated differently from the objects of classical mechanics. Particles are described by wave functions that evolve according to the Schrödinger equation. Non-stationary wave functions are linear combinations of the eigenstates of the system, a phenomenon described by the superposition principle. When a quantum system interacts with a classical system—i.e. when an observable is measured—the system appears to collapse to a random eigenstate of that observable from a classical vantage point.

If collapse is truly random, then no process or algorithm can deterministically predict its outcome. This provided Penrose with a candidate for the physical basis of the non-computable process that he hypothesized to exist in the brain. However, he disliked the random nature of environmentally-induced collapse, as randomness was not a promising basis for mathematical understanding. Penrose proposed that isolated systems may still undergo a new form of wave function collapse, which he called objective reduction (OR).

Details

Penrose sought to reconcile general relativity and quantum theory using his own ideas about the possible structure of spacetime. He suggested that at the Planck scale curved spacetime is not continuous, but discrete. Penrose postulated that each separated quantum superposition has its own piece of spacetime curvature, a blister in spacetime. Penrose suggests that gravity exerts a force on these spacetime blisters, which become unstable above the Planck scale of and collapse to just one of the possible states. The rough threshold for OR is given by Penrose's indeterminacy principle:

where:

- is the time until OR occurs,

- is the gravitational self-energy or the degree of spacetime separation given by the superpositioned mass, and

- is the reduced Planck constant.

Thus, the greater the mass-energy of the object, the faster it will undergo OR and vice versa. Atomic-level superpositions would require 10 million years to reach OR threshold, while an isolated 1 kilogram object would reach OR threshold in 10^{-37}s. Objects somewhere between these two scales could collapse on a timescale relevant to neural processing.

An essential feature of Penrose's theory is that the choice of states when objective reduction occurs is selected neither randomly (as are choices following wave function collapse) nor algorithmically. Rather, states are selected by a "non-computable" influence embedded in the Planck scale of spacetime geometry. Penrose claimed that such information is Platonic, representing pure mathematical truth, aesthetic and ethical values at the Planck scale. This relates to Penrose's ideas concerning the three worlds: physical, mental, and the Platonic mathematical world. In his theory, the Platonic world corresponds to the geometry of fundamental spacetime that is claimed to support non-computational thinking.

No evidence supports Penrose's objective reduction, but the theory is considered testable and the FELIX (experiment) has been suggested to evaluate and measure the objective criterion.

In August 2013, Penrose and Hameroff reported that the experiments had been carried out by Bandyopadhyay et al., supporting Penrose's theory on six out of his twenty theses, while invalidating none of the others. They subsequently responded to critiques, including a 2013 critique from Reimers' group.

The creation of the Orch-OR Model

Penrose and Hameroff initially developed their ideas quite separately from one another, and it was only in the 1990s that they cooperated to produce the Orch-OR theory. Penrose came to the problem from the view point of mathematics and in particular Gödel's theorem, while Hameroff approached it from a career in cancer research and anesthesia that had given him an interest in brain structures. Specifically, when Penrose wrote his first consciousness book, *The Emperor's New Mind* in 1989, he lacked a detailed proposal for how such quantum processes could be implemented in the brain. Subsequently, Hameroff read *The Emperor's New Mind* and suggested to Penrose that certain structures within brain cells (neurons) were suitable candidate sites for quantum processing and ultimately for consciousness. The Orch-OR theory arose from the cooperation of these two scientists, and was developed in Penrose's second consciousness book *Shadows of the Mind* (1994).

Hameroff's contribution to the theory derived from studying brain cells. His interest centered on the cytoskeleton, which provides an internal supportive structure for neurons, and particularly on the microtubules, which are the most important component of the cytoskeleton. As neuroscience has progressed, the role of the cytoskeleton and microtubules has assumed greater importance. In addition to providing structural support, microtubule functions include axoplasmic transport and control of the cell's movement, growth and shape.

Microtubule Condensates

Hameroff proposed that microtubules were suitable candidates for quantum processing. Microtubules are made up of tubulinprotein subunits. The tubulin protein dimers of the microtubules have hydrophobic pockets that may contain delocalized π electrons. Tubulin has other, smaller non-polar regions, for example 8 tryptophans per tubulin, which contain π electron-rich indole rings distributed throughout tubulin with separations of roughly 2 nm. Hameroff claims that this is close enough for the tubulin π electrons to become quantum entangled. During entanglement, particle states become inseparably correlated.

Hameroff originally suggested the tubulin-subunit electrons would form a Bose–Einstein condensate,. He then proposed a Frohlich condensate, a hypothetical coherent oscillation of dipolar molecules. However, this too was rejected by Reimers' group. Hameroff then responded to Reimers. "Reimers et al have most definitely NOT shown that strong or coherent Frohlich condensation in microtubules is unfeasible. The model microtubule on which they base their Hamiltonian is not a microtubule structure, but a simple linear chain of oscillators." Hameroff reasoned that such condensate behavior would magnify nanoscopic quantum effects to have large scale influences in the brain.

Hameroff proposed that condensates in microtubules in one neuron can link with microtubule condensates in other neurons and glial cells via the gap junctions of electrical synapses. Hameroff proposed that the gap between the cells is sufficiently small that quantum objects can tunnel across it, allowing them to extend across a large area of the brain. He further postulated that the action of this large-scale quantum activity is the source of 40 Hz gamma waves, building upon the much less controversial theory that gap junctions are related to the gamma oscillation.

Consequences

The Orch-OR theory combines the Penrose–Lucas argument with Hameroff's hypothesis on quantum processing in microtubules. It proposes that when condensates in the brain undergo an objective wave function reduction, their collapse connects non-computational decision-making to experiences embedded in spacetime's fundamental geometry.

The theory further proposes that the microtubules both influence and are influenced by the conventional activity at the synapses between neurons.

In 1998 Hameroff made 8 probable assumptions and 20 testable predictions to back his proposal.

In January 2014 Hameroff and Penrose announced that the discovery of quantum vibrations in microtubules by Anirban Bandyopadhyay of the National Institute for Materials Science in Japan provides favorable evidence towards the Orch-OR hypothesis.

Criticism

The Orch-OR theory was criticized by scientists who considered it to be a poor model of brain physiology.

Penrose–Lucas Argument

The Penrose–Lucas argument about the implications of Gödel's incompleteness theorem for computational theories of human intelligence was criticized by mathematicians, computer scientists, and philosophers, and the consensus among experts in these fields is that the argument fails, with different authors attacking different aspects of the argument.

LaForte pointed out that in order to know the truth of an unprovable Gödel sentence, one must already know the formal system is consistent. Referencing Benacerraf, he then demonstrated that humans cannot prove that they are consistent, and in all likelihood human brains are inconsistent. He pointed to contradictions within Penrose's own writings as examples. Similarly, Minsky argued that because humans can believe false ideas to be true, human mathematical understanding need not be consistent and consciousness may easily have a deterministic basis.

Feferman faulted detailed points in Penrose's second book, *Shadows of the Mind*. He argued that mathematicians do not progress by mechanistic search through proofs, but by trial-and-error reasoning, insight and inspiration, and that machines do not share this approach with humans. He pointed out that everyday mathematics can be formalized. He also rejected Penrose's Platonism.

Searle criticized Penrose's appeal to Gödel as resting on the fallacy that all computational algorithms must be capable of mathematical description. As a counter-example, Searle cited the assignment of license plate numbers to specific vehicle identification numbers, as part of vehicle registration. According to Searle, no mathematical function can be used to connect a known VIN with its LPN, but the process of assignment is quite simple—namely, "first come, first served"—and can be performed entirely by a computer. However, since an algorithm (as defined in the Oxford American Dictionary) is a 'set of rules to be followed in calculations or problem-solving operations', the assignment of LPN to a VIN is not an algorithm as such, merely the use of a database in which every VIN has a corresponding LPN. No algorithm could arbitrarily 'compute' database assignments. Thus, Searle's counter-example does not describe a computational algorithm that is not mathematically describable.

Decoherence in Living Organisms

In 2000 Tegmark claimed that any quantum coherent system in the brain would undergo wave function collapse due to environmental interaction long before it could influence neural processes (the *"warm, wet and noisy"* argument, as it was later came to be known). He determined the decoherence timescale of microtubule entanglement at brain temperatures to be on the order of femtoseconds, far too brief for neural processing. Other scientists sided with Tegmark's analysis, insisting that quantum coherence does not play, or does not need to play any major role in neurophysiology.

In response to Tegmark's claims, Hagan, Tuszynski and Hameroff claimed that Tegmark did not address the Orch-OR model, but instead a model of his own construction. This involved superpositions of quanta separated by 24 nm rather than the much smaller separations stipulated for Orch-OR. As a result, Hameroff's group claimed a decoherence time seven orders of magnitude greater than Tegmark's, although still far below 25 ms. Hameroff's group also suggested that the Debye layer of counterions could screen thermal fluctuations, and that the surrounding actin gel might enhance the ordering of water, further screening noise. They also suggested that incoherent metabolic energy could further order water, and finally that the configuration of the microtubule lattice might be suitable for quantum error correction, a means of resisting quantum decoherence.

Since the 90's numerous counter-observations to the "warm, wet and noisy" argument existed at ambient temperatures, *in vitro* and *in vivo* (i.e. photosynthesis, bird navigation). For example, Harvard researchers achieved quantum states lasting for 2 sec at room temperatures using diamonds. Plants routinely use quantum-coherent electron transport at ambient temperatures in photosynthesis. In 2014, researchers used theoretical quantum biophysics and computer simulations to analyze quantum coherence among tryptophan π resonance rings in tubulin. They claimed that quantum dipole coupling among tryptophan π resonance clouds, mediated by exciton hopping or Forster resonance energy transfer (FRET) across the tubulin protein are plausible.

In 2009, Reimers *et al.* and McKemmish *et al.*, published critical assessments. Earlier versions of the theory had required tubulin-electrons to form either Bose–Einsteins or Frohlich condensates, and the Reimers group claimed that these were experimentally unfounded. Additionally they claimed that microtubules could only support 'weak' 8 MHz coherence. The first argument was voided by revisions of the theory that described dipole oscillations due to London forces and possibly due to magnetic and/or nuclear spin cloud formations. On the second issue the theory was retrofitted so that 8 MHz coherence is sufficient to support the whole Orch-OR hypothesis.

McKemmish *et al.* made two claims: that aromatic molecules cannot switch states because they are delocalised; and that changes in tubulin protein-conformation driven by GTP conversion would result in a prohibitive energy requirement. Hameroff and Penrose responded to the first claim by stating that they were referring to the behaviour of two or more electron clouds, inherently non-localised. For the second claim they stated that no GTP conversion is needed since (in that version of the theory) the conformation-switching is not necessary, replaced by oscillation due to the London forces produced by the electron cloud dipole states.

Neuron Cell Biology

Hameroff proposed that microtubule coherence reaches the synapses via dendritic lamellar bodies (DLBs), where it could influence synaptic firing and be transmitted across the synaptic cleft. De Zeeuw *et al.* then proved this to be impossible, by showing that DLBs are located micrometers away from gap junctions. Bandyopadhyay *et. al.* speculated that this issue might be resolved if their notion of wireless transmission of information globally across the entire brain is proven. Hameroff and Penrose doubt whether such a wireless transmission would be capable of transmitting superimposed quantum-states.

Hameroff's 1998 hypothesis required that cortical dendrites contain primarily 'A' lattice microtubules, but in 1994 Kikkawa *et al.* showed that all *in vivo* microtubules have a 'B' lattice and a seam. Then Bandyopadhyay showed that microtubules can change their structure from B-lattice to A-lattice as part of the processing of information, and that tubulin in microtubules exists in multiple states.

Orch-OR also required gap junctions between neurons and glial cells, yet Binmöller *et. al.* proved that these don't exist. But recent research supports the evidence between neurons and astrocytes.

Hameroff speculated that visual photons in the retina are detected directly by the cones and rods instead of decohering and subsequently connect with the retinal glia cells via gap junctions, but this too was falsified.

Other biology-based criticisms have been offered. Papers by Georgiev point to problems with Hameroff's proposals, including a lack of explanation for the probabilistic firing of axonal synapses and an error in the calculated number of the tubulin dimers per cortical neuron. Hameroff insisted in a 2013 interview that those falsifications were invalid.

Vibration Theory of Olfaction

The Vibration theory of smell proposes that a molecule's smell character is due to its vibrational frequency in the infrared range. This controversial theory is an alternative to the more widely accepted shape theory of olfaction, which proposes that a molecule's smell character is due to its molecular size, molecular shape and functional groups.

Introduction

The current vibration theory has recently been called the "swipe card" model, in contrast with "lock and key" models based on shape theory. As proposed by Luca Turin, the odorant molecule must first fit in the receptor's binding site. Then it must have a vibrational energy mode compatible with the difference in energies between two energy levels on the receptor, so electrons can travel through the molecule via inelastic electron tunneling, triggering the signal transduction pathway. The vibration theory is discussed in a popular, but controversial book by Chandler Burr.

The odor character is encoded in the ratio of activities of receptors tuned to different vibration frequencies, in the same way that color is encoded in the ratio of activities of cone cell receptors tuned to different frequencies of light. An important difference, though, is that the odorant has to be able to become resident in the receptor for a response to be generated. The time an odorant resides in a receptor depends on how strongly it binds, which in turn determines the strength of the response; the odor intensity is thus governed by a similar mechanism to the "lock and key" model. For a pure vibrational theory, the differing odors of enantiomers, which possess identical vibrations, cannot be explained. However, once the link between receptor response and duration of the residence of the odorant in the receptor is recognised, differences in odor between enantiomers can be understood: molecules with different handedness may spend different amounts of time in a given receptor, and so initiate responses of different intensities.

Some studies support vibration theory while others challenge its findings.

Major Proponents and History

The theory was first proposed by Malcolm Dyson in 1928 and expanded by Robert H. Wright in 1954, after which it was largely abandoned in favor of the competing shape theory. A 1996 paper by Luca Turin revived the theory by proposing a mechanism, speculating that the G-protein-coupled receptors discovered by Linda Buck and Richard Axel were actually measuring molecular vibrations using inelastic electron tunneling as Turin claimed, rather than responding to molecular keys fitting molecular locks, working by shape alone. In 2007 a Physical Review Letters paper by Marshall Stoneham and colleagues at University College London and Imperial College London showed that Turin's proposed mechanism was consistent with known physics and coined the expression "swipe card model" to describe it. A PNAS paper in 2011 by Turin, Efthimios Skoulakis, and colleagues at MIT and the Alexander Fleming Biomedical Sciences Research Center reported fly behavioral experiments consistent with a vibrational theory of smell. The theory remains controversial.

Support

Isotope Effects

A major prediction of Turin's theory is the isotope effect: that the normal and deuterated versions of a compound should smell different, although they have the same shape. A 2001 study by Haffenden et al. showed humans able to distinguish benzaldehyde from its deuterated version. However, this study has been criticized for lacking double-blind controls to eliminate bias and because it used an anomalous version of the duo-trio test. In addition, tests with animals have shown fish and insects able to distinguish isotopes by smell.

It should be noted, however, that deuteration changes the heats of adsorption and the boiling and freezing points of molecules (boiling points: 100.0 °C for H_2O vs. 101.42 °C for D_2O; melting points: 0.0 °C for H_2O, 3.82 °C for D_2O), pK_a (i.e., dissociation constant: 9.71×10^{-15} for H_2O vs. 1.95×10^{-15} for D_2O, cf. Heavy water) and the strength of hydrogen bonding. Such isotope effects are exceedingly common, and so it is well known that deuterium substitution will indeed change the binding constants of molecules to protein receptors. Any binding interaction of an odorant molecule with an olfactory receptor will therefore be likely to show some isotope effect upon deuteration, and the observation of an isotope effect in no way argues exclusively for a vibrational theory of olfaction.

A study published in 2011 by Franco, Turin, Mershin and Skoulakis shows both that flies can smell deuterium, and that to flies, a carbon-deuterium bond smells like a nitrile, which has a similar vibration. The study reports that drosophila melanogaster (fruit fly), which is ordinarily attracted to acetophenone, spontaneously dislikes deuterated acetophenone. This dislike increases with the number of deuteriums. (Flies genetically altered to lack smell receptors could not tell the difference.) Flies could also be trained by electric shocks either to avoid the deuterated molecule or to prefer it to the normal one. When these trained flies were then presented with a completely new and unrelated choice of normal vs. deuterated odorants, they avoided or preferred deuterium as with the previous pair. This suggested that flies were able to smell deuterium regardless of the rest of the molecule. To determine whether this deuterium smell was actually due to vibrations of the carbon-deuterium (C-D) bond or to some unforeseen effect of isotopes, the researchers looked to nitriles, which have a similar vibration to the C-D bond. Flies trained to avoid deuterium and asked to choose between a nitrile and its non-nitrile counterpart did avoid the nitrile, lending support to the idea that the flies are smelling vibrations. Further isotope smell studies are under way in fruit flies and dogs.

Explaining Differences in Stereoisomer Scents

Carvone presented a perplexing situation to vibration theory. Carvone has two isomers, which have identical vibrations, yet one smells like mint and the other like caraway (for which the compound is named).

An experiment by Turin filmed by the 1995 BBC Horizon documentary "A Code in the Nose" consisted of mixing the mint isomer with butanone, on the theory that the shape of the G-protein-coupled receptor prevented the carbonyl group in the mint isomer from being detected by the "biological spectroscope". The experiment succeeded with the trained perfumers used as subjects, who perceived that a mixture of 60% butanone and 40% mint carvone smelled like caraway.

The Sulfurous Smell of Boranes

According to Turin's original paper in the journal *Chemical Senses*, the well documented smell of borane compounds is intensely sulfurous, though these molecules contain no sulfur. He proposes to explain this by the similarity in frequency between the vibration of the B-H bond and the S-H bond.

Consistency with Physics

Biophysical simulations published in Physical Review Letters in 2006 suggest that Turin's proposal is viable from a physics standpoint. However, Block et al. in their 2015 paper in Proceedings of the National Academy of Sciences indicate that their theoretical analysis shows that "the proposed electron transfer mechanism of the vibrational frequencies of odorants could be easily suppressed by quantum effects of nonodorant molecular vibrational modes."

Correlating Odor to Vibration

A 2004 paper published in the journal *Organic Biomolecular Chemistry* by Takane and Mitchell shows that odor descriptions in the olfaction literature correlate with EVA descriptors, which loosely correspond to the vibrational spectrum, better than with descriptors based on the two dimensional connectivity of the molecule. The study did not consider molecular shape.

Lack of Antagonists

Turin points out that traditional lock-and-key receptor interactions deal with agonists, which increase the receptor's time spent in the active state, and antagonists, which increase the time spent in the inactive state. In other words, some ligands tend to turn the receptor on and some tend to turn it off. As an argument against the traditional lock-and-key theory of smell, no olfactory antagonists have yet been found until recently.

In 2004, a Japanese research group published that an oxidation product of isoeugenol is able to antagonize, or prevent, mice olfactory receptor response to isoeugenol.

Additional Challenges to Shape Theory

- Similarly shaped molecules with different molecular vibrations have different smells (metallocene experiment and deuterium replacement of molecular hydrogen). However this challenge is contrary to the results obtained with silicon analogues of bourgeonal and lilial, which despite their differences in molecular vibrations have similar smells and similarly activate the most responsive human receptor, hOR17-4, and with studies showing that the human musk receptor OR5AN1 responds identically to deuterated and non-deuterated musks.

- Differently shaped molecules with similar molecular vibrations have similar smells (replacement of carbon double bonds by sulfur atoms and the disparate shaped amber odorants)

- Hiding functional groups does not hide the group's characteristic odor. However this is not always the case, since *ortho*-substituted arylisonitriles and thiophenols have far less offensive odors than the parent compounds.

Challenges to Vibration Theory

Three predictions by Luca Turin on the nature of smell, using concepts of vibration theory, were addressed by experimental tests published in Nature Neuroscience in 2004 by Vosshall and Keller. The study failed to support the prediction that isotopes should smell different, with untrained human subjects unable to distinguish acetophenone from its deuterated counterpart. This study also pointed to experimental design flaws in the earlier study by Haffenden. In addition, Turin's description of the odor of long-chain aldehydes as alternately (1) dominantly waxy and faintly citrus and (2) dominantly citrus and faintly waxy was not supported by tests on untrained subjects, despite anecdotal support from fragrance industry professionals who work regularly with these materials. Vosshall and Keller also presented a mixture of guaiacol and benzaldehyde to subjects, to test Turin's theory that the mixture should smell of vanillin. Vosshall and Keller's data did not support Turin's prediction. However, Vosshall says these tests do not disprove the vibration theory.

In response to the 2011 PNAS study on flies, Vosshall acknowledged that flies could smell isotopes but called the conclusion that smell was based on vibrations an "overinterpretation" and expressed skepticism about using flies to test a mechanism originally ascribed to human receptors. For the theory to be confirmed, Vosshall stated there must be further studies on mammalian receptors. Bill Hansson, an insect olfaction specialist, raised the question of whether deuterium could affect hydrogen bonds between the odorant and receptor.

In 2013, Turin and coworkers confirmed Vosshall and Keller's experiments showing that even trained human subjects were unable to distinguish acetophenone from its deuterated counterpart. At the same time Turin and coworkers reported that human volunteers were able to distinguish cyclopentadecanone from its fully deuterated analog. To account for the different results seen with acetophenone and cyclopentadecanone, Turin and coworkers assert that "there must be many C-H bonds before they are detectable by smell. In contrast to acetophenone which contains only 8 hydrogens, cyclopentadecanone has 28. This results in more than 3 times the number of vibrational modes involving hydrogens than in acetophenone, and this is likely essential for detecting the difference between isotopomers." Turin and coworkers provide no quantum mechanical justification for this latter assertion.

Vosshall, in commenting on Turin's work, notes that "the olfactory membranes are loaded with enzymes that can metabolise odorants, changing their chemical identity and perceived odour. Deuterated molecules would be poor substrates for such enzymes, leading to a chemical difference in what the subjects are testing. Ultimately, any attempt to prove the vibrational theory of olfaction should concentrate on actual mechanisms at the level of the receptor, not on indirect psychophysical testing." Richard Axel co-recipient of the 2004 Nobel prize for physiology for his work on olfaction, expresses a similar sentiment, indicating that Turin's work "would not resolve the debate - only a microscopic look at the receptors in the nose would finally show what is at work. Until somebody really sits down and seriously addresses the mechanism and not inferences from the mechanism... it doesn't seem a useful endeavour to use behavioural responses as an argument."

In response to the 2013 paper on cyclopentadecanone, Block et al. report that the human musk-recognizing receptor, OR5AN1, identified using a heterologous olfactory receptor expression system and robustly responding to cyclopentadecanone and muscone (which has 30 hydrogens), fails to distinguish isotopomers of these compounds in vitro. Furthermore, the mouse (methylthio)methanethiol-recognizing receptor, MOR244-3, as well as other selected human and mouse olfactory receptors, responded similarly to normal, deuterated, and carbon-13 isotopomers of their respective ligands, paralleling results found with the musk receptor OR5AN1. Based on these findings, the authors conclude that the proposed vibration theory does not apply to the human musk receptor OR5AN1, mouse thiol receptor MOR244-3, or other olfactory receptors examined. Additionally, theoretical analysis by the authors shows that the proposed electron transfer mechanism of the vibrational frequencies of odorants could be easily suppressed by quantum effects of nonodorant molecular vibrational modes. The authors conclude: "These and other concerns about electron transfer at olfactory receptors, together with our extensive experimental data, argue against the plausibility of the vibration theory."

In commenting on this work, Vosshall writes "In PNAS, Block et al.…. shift the "shape vs. vibration" debate from olfactory psychophysics to the biophysics of the ORs themselves. The authors mount a sophisticated multidisciplinary attack on the central tenets of the vibration theory using synthetic organic chemistry, heterologous expression of olfactory receptors, and theoretical considerations to find no evidence to support the vibration theory of smell." While Turin comments that Block used "cells in a dish rather than within whole organisms" and that "expressing an olfactory receptor in human embryonic kidney cells doesn't adequately reconstitute the complex nature of olfaction…", Vosshall responds "Embryonic kidney cells are not identical to the cells in the nose .. but if you are looking at receptors, it's the best system in the world." In a Letter to the Editor of PNAS, Turin et al. raise concerns about Block et al. and Block et al. respond.

Recently, Saberi and Allaei have suggested that a functional relationship exists between molecular volume and the olfactory neural response. The molecular volume is an important factor, but it is not the only factor that determines the response of ORNs. The binding affinity of an odorant-receptor pair is affected by their relative sizes. The maximum affinity can be attained when the molecular volume of an odorant matches the volume of the binding pocket.

Biomagnetism

Biomagnetism is the phenomenon of magnetic fields*produced* by living organisms; it is a subset of bioelectromagnetism. In contrast, organisms' use of magnetism in navigation is magnetoception and the study of the magnetic fields' *effects* on organisms is *magnetobiology*. (The word biomagnetism has also been used loosely to include magnetobiology, further encompassing almost any combination of the words magnetism, cosmology, and biology, such as "magnetoastrobiology".)

The origin of the word biomagnetism is unclear, but seems to have appeared several hundred years ago, linked to the expression "animal magnetism." The present scientific definition took form in the 1970s, when an increasing number of researchers began to measure the magnetic fields produced by the human body. The first valid measurement was actually made in 1963, but the field of research began to expand only after a low-noise technique was developed in 1970. Today the community of biomagnetic researchers does not have a formal organization, but international conferences are held every two years, with about 600 attendees. Most conference activity centers on the MEG (magnetoencephalogram), the measurement of the magnetic field of the brain.

Prominent Researchers

- David Cohen

- John Wikswo

- Samuel Williamson

References

- Penrose, Roger (1989). The Emperor's New Mind: Concerning Computers, Minds and The Laws of Physics. Oxford University Press. p. 480. ISBN 0-19-851973-7.

- Penrose, Roger (1989). Shadows of the Mind: A Search for the Missing Science of Consciousness. Oxford University Press. p. 457. ISBN 0-19-853978-9.

- Burr, C. (2003). The Emperor of Scent: A Story of Perfume, Obsession, and the Last Mystery of the Senses. Random House. ISBN 0-375-50797-3.

- "A Biomagnetic Sensory Mechanism Based on Magnetic Field Modulated Coherent Electron Spin Motion : Zeitschrift für Physikalische Chemie". www.degruyter.com. Retrieved 2015-12-01.

- "A Biomagnetic Sensory Mechanism Based on Magnetic Field Modulated Coherent Electron Spin Motion : Zeitschrift für Physikalische Chemie". Retrieved 2015-08-20.

- "Fox 'rangefinder' sense expands the magnetic menagerie". blogs.nature.com. Nature Publishing Group / Macmillan. Retrieved 6 June 2014.

- Penrose, Roger (2014). "On the Gravitization of Quantum Mechanics 1: Quantum State Reduction". Foundations of Physics. 44: 557–575. Bibcode:2014FoPh...44..557P. doi:10.1007/s10701-013-9770-0. Retrieved 12 June 2016.

- HODGKIN; HUXLEY (1952). "A quantitative description of membrane current and its application to conduction and excitation in nerve."(PDF). The Journal of Physiology. MEDLINE. 117 (4): 500–544. Retrieved 17 March 2014.

- Hameroff, Stuart; Penrose, Roger (March 2014). "Consciousness in the universe: A review of the 'Orch OR' theory". Physics of Life Reviews. Elsevier. 11 (1): 39–78. doi:10.1016/j.plrev.2013.08.002. PMID 24070914. Retrieved 16 March 2014.

- Natalie Wolchover (31 October 2013). "Physicists Eye Quantum-Gravity Interface". Quanta Magazine (Article). Simons Foundation. Retrieved 19 March 2014.

Related Aspects of Biophysics

6

The related aspects of biophysics are known as lipid polymorphism, biophysical profile, medical physics and cell biophysics. Lipid polymorphism is the capability of lipids to aggregate in different ways to form structures. The topics discussed in the chapter are of great importance to broaden the existing knowledge on biophysics.

Lipid Polymorphism

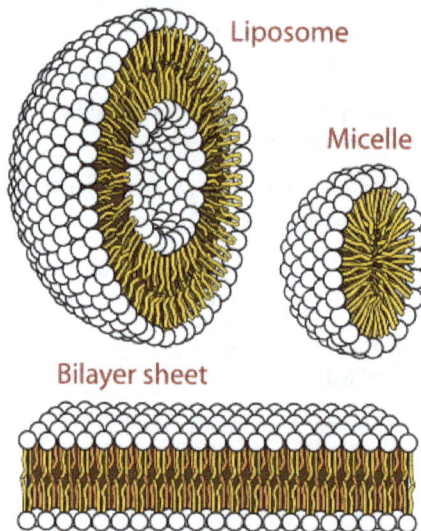

Cross Section view of the structures that can be formed by phospholipids in aqueous solutions

Polymorphism in biophysics is the ability of lipids to aggregate in a variety of ways, giving rise to structures of different shapes, known as "phases". This can be in the form of spheres of lipid molecules (micelles), pairs of layers that face one another (lamellar phase, observed in biological systems as a lipid bilayer), a tubular arrangement (hexagonal), or various cubic phases (Fd3m, Im3m, Ia3m, Pn3m, and Pm3m being those discovered so far). More complicated aggregations have also been observed, such as rhombohedral, tetragonal and orthorhombic phases.

It forms an important part of current academic research in the fields of membrane biophysics (polymorphism), biochemistry (biological impact) and organic chemistry (synthesis).

Determination of the topology of a lipid system is possible by a number of methods, the most reliable of which is x-ray diffraction. This uses a beam of x-rays that are scattered by the sample, giving a diffraction pattern as a set of rings. The ratio of the distances of these rings from the central point indicates which phase(s) are present.

The structural phase of the aggregation is influenced by the ratio of lipids present, temperature, hydration, pressure and ionic strength (and type).

Hexagonal Phases in the Lipid Polymorphism

In lipid polymorphism, if the packing ratio of lipids is greater or less than one, lipid membranes can form two separate hexagonal phases, or nonlamellar phases, in which long, tubular aggregates form according to the environment in which the lipid is introduced.

Hexagonal I Phase (H$_I$)

This phase is favored in detergent-in-water solutions and has a packing ratio of less than one. The micellar population in a detergent/water mixture cannot increase without limit as the detergent to water ratio increases. In the presence of low amounts of water, lipids that would normally form micelles will form larger aggregates in the form of micellar tubules in order to satisfy the requirements of the hydrophobic effect. These aggregates can be thought of as micelles that are fused together. These tubes have the polar head groups facing out, and the hydrophobic hydrocarbon chains facing the interior. This phase is only seen under unique, specialized conditions, and most likely is not relevant for biological membranes.

Hexagonal II Phase (H$_{II}$)

Lipid molecules in the HII phase pack inversely to the packing observed in the hexagonal I phase described above. This phase has the polar head groups on the inside and the hydrophobic, hydrocarbon tails on the outside in solution. The packing ratio for this phase is less than one, which is synonymous with an inverse cone packing.

Extended arrays of long tubes will form (as in the hexagonal I phase), but because of the way the polar head groups pack, the tubes take the shape of aqueous channels. These arrays can stack together like pipes. This way of packing may leave a finite hydrophobic surface in contact with water on the outside of the array. However, the otherwise energetically favorable packing apparently stabilizes this phase as a whole. It is also possible that an outer monolayer of lipid coats the surface of the collection of tubes to protect the hydrophobic surface from interaction with the aqueous phase.

It is suggested that this phase is formed by lipids in solution in order to compensate for the hydrophobic effect. The tight packing of the lipid head groups reduces their contact with the aqueous phase. This, in turn, reduces the amount of ordered, but unbound water molecules. The most common lipids that form this phase include phospatidyleth-anolamine (PE), when it has unsaturated hydrocarbon chains. Diphosphatidylglycerol (DPG, otherwise known as cardiolipin) in the presence of calcium is also capable of forming this phase.

Techniques for Detection

There are several techniques used to map out which phase is present during perturbations done on the lipid. These perturbations include pH changes, temperature changes, pressure changes, volume changes, etc.

The most common technique used to study phospholipid phase presence is phosphorus nuclear magnetic resonance (31P NMR). In this technique, different and unique powder diffraction patterns are observed for lamellar, hexagonal, and isotropic phases. Other techniques that are used and do offer definitive evidence of existence of lamellar and hexagonal phases include freeze-fracture electron microscopy, X-Ray diffraction, differential scanning calorimetry (DSC), and deuterium nuclear magnetic resonance (2H NMR).

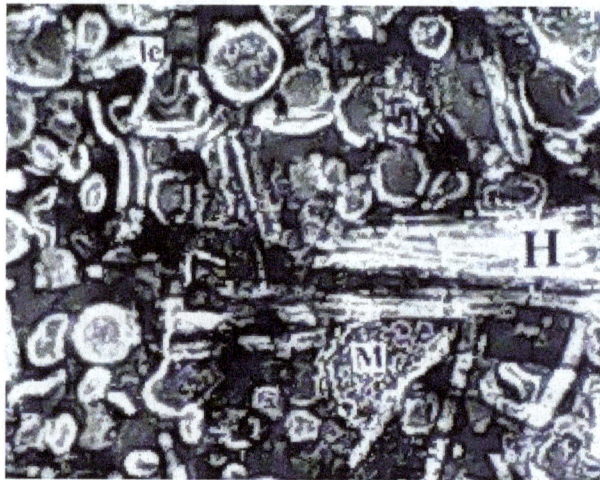

Inverted hexagonal H-II phase (H), inverted spherical micellar phase (M), lamellar liposomal phase (le) structures in cold-exposed aqueous dispersions of total lipid extract of spinach thylakoid membranes, studied by negative staining (2% phosphotungstic acid) transmission electron microscopy.

Additionally, negative staining transmission electron microscopy has been shown as a useful tool to study lipid bilayer phase behavior and polymorphism into *lamellar phase, micellar, unilamellar liposome, and hexagonal aqueous-lipid structures*, in aqueous dispersions of membrane lipids. As water-soluble negative stain is excluded from the hydrophobic part (fatty acyl chains) of lipid aggregates, the hydrophilic head-group portions of the lipid aggregates stain dark and clearly mark the outlines of the lipid aggregates.

Biophysical Profile

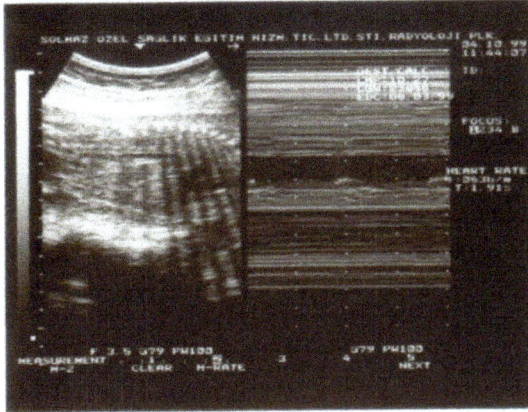

Ultrasound

A biophysical profile (BPP) is a prenatalultrasound evaluation of fetal well-being involving a scoring system, with the score being termed Manning's score. It is often done when a non-stress test (NST) is non reactive, or for other obstetrical indications.

The "modified biophysical profile" consists of the NST and amniotic fluid index only.

Procedure

The BPP has 5 components: 4 ultrasound (US) assessments and an NST. The NST evaluates fetal heart rate and response to fetal movement. The five discrete biophysical variables:

1. Fetal heart rate

2. Fetal breathing

3. Fetal movement

4. Fetal tone

5. Amniotic fluid volume

Parameter	Normal (2 points)	Abnormal (0 points)
NST/Reactive FHR	At least two accelerations in 20 minutes	Less than two accelerations to satisfy the test in 20 minutes
US: Fetal breathing movements	At least one episode of > 30s or >20s in 30 minutes	None or less than 30s or 20s
US: Fetal activity / gross body movements	At least three or two movements of the torso or limbs	Less than three or two movements

US: Fetal muscle tone	At least one episode of active bending and straightening of the limb or trunk	No movements or movements slow and incomplete
US: Qualitative AFV/AFI	At least one vertical pocket> 2 cm or more in the vertical axis	Largest vertical pocket</=2 cm

Use of vibroacoustic stimulation to accelerate evaluation has been described.

Interpretation

Each assessment is graded either 0 or 2 points, and then added up to yield a number between 0 and 10. A BPP of 8 or 10 is generally considered reassuring. A BPP normally is not performed before the second half of a pregnancy, since fetal movements do not occur in the first half. The presence of these biophysical variables implies absence of significant central nervous system hypoxemia/acidemia at the time of testing. By comparison, a compromised fetus typically exhibits loss of accelerations of the fetal heart rate (FHR), decreased body movement and breathing, hypotonia, and, less acutely, decreased amniotic fluid volume.

Recommended management based on the biophysical profile	
BPP	**Recommended management**
≤2	• Labor induction
4	• Labor induction if gestational age>32 weeks • Repeating test same day if <32 weeks, then delivery if BPP <6
6	• Labor induction if >36 weeks if favorable cervix and normal AFI • Repeating test in 24 hours if <36 weeks and cervix unfavorable; then delivery if BPP <6, and follow-up if >6
8	• Labor induction if presence of oligohydramnios

Medical Physics

Medical physics (also called biomedical physics, medical biophysics or applied physics in medicine) is, generally speaking, the application of physics concepts, theories and methods to medicine or healthcare. Medical physics departments may be found in hospitals or universities.

In the case of hospital work, the term 'Medical Physicist' is the title of a specific healthcare profession with a specific mission statement. Such Medical Physicists are often found in the following healthcare specialties: diagnostic and intervention radiology (also known as medical imaging), nuclear medicine, and radiation oncology (also known as radiotherapy). However, areas of specialty are widely varied in scope and breadth, e.g. clinical physiology (also known as physiological measurement, several countries), neurophysiology (Finland), radiation protection (many countries), and audiology (Netherlands).

University departments are of two types. The first type are mainly concerned with preparing students for a career as a hospital medical physicist and research focuses on improving the practice of the profession. A second type (increasingly called 'biomedical physics') has a much wider scope and may include research in any applications of physics to medicine from the study of biomolecular structure to microscopy and nanomedicine. For example, physicist Richard Feynman theorized about the future of nanomedicine. He wrote about the idea of a *medical* use for biological machines. Feynman and Albert Hibbs suggested that certain repair machines might one day be reduced in size to the point that it would be possible to (as Feynman put it) "swallow the doctor". The idea was discussed in Feynman's 1959 essay *There's Plenty of Room at the Bottom*.

Mission statement of Medical Physicists

In the case of hospital medical physics departments, the mission statement is the following:

"Medical Physicists will contribute to maintaining and improving the quality, safety and cost-effectiveness of healthcare services through patient-oriented activities requiring expert action, involvement or advice regarding the specification, selection, acceptance testing, commissioning, quality assurance/control and optimised clinical use of medical devices and regarding patient risks and protection from associated physical agents (e.g., x-rays, electromagnetic fields, laser light, radionuclides) including the prevention of unintended or accidental exposures; all activities will be based on current best evidence or own scientific research when the available evidence is not sufficient. The scope includes risks to volunteers in biomedical research, carers and comforters. The scope often includes risks to workers and public particularly when these impact patient risk"

The term "physical agents" refers to ionising and non-ionising electromagnetic radiations, static electric and magnetic fields, ultrasound, laser light and any other Physical Agent associated with medical e.g., x-rays in computerised tomography (CT), gamma rays/radionuclides in nuclear medicine, magnetic fields and radio-frequencies in magnetic resonance imaging (MRI), ultrasound in ultrasound imaging and Doppler measurements etc.

This mission includes the following 11 key activities:

1. Scientific problem solving service: Comprehensive problem solving service involving recognition of less than optimal performance or optimised use of medical devices, identification and elimination of possible causes or misuse, and

confirmation that proposed solutions have restored device performance and use to acceptable status. All activities are to be based on current best scientific evidence or own research when the available evidence is not sufficient.

2. Dosimetry measurements: Measurement of doses suffered by patients, volunteers in biomedical research, carers, comforters and persons subjected to non-medical imaging exposures (e.g., for legal or employment purposes); selection, calibration and maintenance of dosimetry related instrumentation; independent checking of dose related quantities provided by dose reporting devices (including software devices); measurement of dose related quantities required as inputs to dose reporting or estimating devices (including software). Measurements to be based on current recommended techniques and protocols. Includes dosimetry of all physical agents.

3. Patient safety/risk management (including volunteers in biomedical research, carers, comforters and persons subjected to non-medical imaging exposures. Surveillance of medical devices and evaluation of clinical protocols to ensure the ongoing protection of patients, volunteers in biomedical research, carers, comforters and persons subjected to non-medical imaging exposures from the deleterious effects of physical agents in accordance with the latest published evidence or own research when the available evidence is not sufficient. Includes the development of risk assessment protocols.

4. Occupational and public safety/risk management (when there is an impact on medical exposure or own safety). Surveillance of medical devices and evaluation of clinical protocols with respect to protection of workers and public when impacting the exposure of patients, volunteers in biomedical research, carers, comforters and persons subjected to non-medical imaging exposures or responsibility with respect to own safety. Includes the development of risk assessment protocols in conjunction with other experts involved in occupational / public risks.

5. Clinical medical device management: Specification, selection, acceptance testing, commissioning and quality assurance/ control of medical devices in accordance with the latest published European or International recommendations and the management and supervision of associated programmes. Testing to be based on current recommended techniques and protocols.

6. Clinical involvement: Carrying out, participating in and supervising everyday radiation protection and quality control procedures to ensure ongoing effective and optimised use of medical radiological devices and including patient specific optimization.

7. Development of service quality and cost-effectiveness: Leading the introduction of new medical radiological devices into clinical service, the introduction of new

medical physics services and participating in the introduction/development of clinical protocols/techniques whilst giving due attention to economic issues.

8. Expert consultancy: Provision of expert advice to outside clients (e.g., clinics with no in-house medical physics expertise).

9. Education of healthcare professionals (including medical physics trainees: Contributing to quality healthcare professional education through knowledge transfer activities concerning the technical-scientific knowledge, skills and competences supporting the clinically effective, safe, evidence-based and economical use of medical radiological devices. Participation in the education of medical physics students and organisation of medical physics residency programmes.

10. Health technology assessment (HTA): Taking responsibility for the physics component of health technology assessments related to medical radiological devices and /or the medical uses of radioactive substances/sources.

11. Innovation: Developing new or modifying existing devices (including software) and protocols for the solution of hitherto unresolved clinical problems.

Medical Biophysics and Biomedical Physics

Some education institutions house departments or programs bearing the title "medical biophysics" or "biomedical physics" or "applied physics in medicine". Generally, these fall into one of two categories: interdisciplinary departments that house biophysics, radiobiology, and medical physics under a single umbrella; and undergraduate programs that prepare students for further study in medical physics, biophysics, or medicine.

Areas of Speciality

The International Organization for Medical Physics (IOMP) - the world's premier professional organization for medical physics with nearly 22,000 members in 84 countries - recognizes main areas of medical physics employment and focus. These are:

Medical Imaging Physics

Para-sagittal MRI of the head in a patient with benign familial macrocephaly.

Medical imaging physics is also known as diagnostic and interventional radiology physics. Clinical (both "in-house" and "consulting") physicists typically deal with areas of testing, optimization, and quality assurance of diagnostic radiology physics areas such as radiographic X-rays, fluoroscopy, mammography, angiography, and computed tomography, as well as non-ionizing radiation modalities such as ultrasound, and MRI. They may also be engaged with radiation protection issues such as radiation exposure monitoring and dosimetry. In addition, many imaging physicists are often also involved with nuclear medicine systems, including single photon emission computed tomography (SPECT) and positron emission tomography (PET). Sometimes, imaging physicists may be engaged in clinical areas, but for research and teaching purposes, such as e.g. quantifying intravascular ultrasound as a possible method of imaging a particular vascular object.

Radiation Therapeutic Physics

Radiation therapeutic physics is also known as radiotherapy physics or radiation oncology physics. The majority of medical physicists currently working in the US, Canada, and some western countries are of this group. A Radiation Therapy physicist typically deals with linear accelerator (Linac) systems and kilovoltage x-ray treatment units on a daily basis, as well as more advanced modalities such as TomoTherapy, gamma knife, cyberknife, proton therapy, and brachytherapy. The academic and research side of therapeutic physics may encompass fields such as boron neutron capture therapy, sealed source radiotherapy, terahertz radiation, high-intensity focused ultrasound (including lithotripsy), optical radiationlasers, ultraviolet etc. including photodynamic therapy, as well as nuclear medicine including unsealed source radiotherapy, and photomedicine, which is the use of light to treat and diagnose disease.

Nuclear Medicine Physics

This is a branch of medicine that uses radiation to provide information about the functioning of a person's specific organs or to treat disease. In most cases, the information is used by physicians to make a quick, accurate diagnosis of the patient's illness. The thyroid, bones, heart, liver and many other organs can be easily imaged, and disorders in their function revealed. In some cases radiation can be used to treat diseased organs, or tumours. Five Nobel Laureates have been intimately involved with the use of radioactive tracers in medicine. Over 10,000 hospitals worldwide use radioisotopes in medicine, and about 90% of the procedures are for diagnosis. The most common radioisotope used in diagnosis is technetium-99, with some 30 million procedures per year, accounting for 80% of all nuclear medicine procedures worldwide. In developed countries (26% of world population) the frequency of diagnostic nuclear medicine is 1.9% per year, and the frequency of therapy with radioisotopes is about one tenth of this. In the USA there are some 18 million nuclear medicine procedures per year among 311 million people, and in Europe about 10 million among 500 million people. In Australia there are about 560,000 per year among 21 million people, 470,000 of these using reactor isotopes. The use of radiopharmaceuticals in diagnosis is growing at over 10% per year. Nuclear medicine was developed in the 1950s by physicians with an endocrine emphasis, initially using iodine-131 to diagnose and then treat thyroid disease. In recent years, specialists have also come from radiology, as dual CT/PET procedures have become established. Computed X-ray tomography (CT) scans and nuclear medicine contribute 36% of the total radiation exposure and 75% of the medical exposure to the

US population, according to a US National Council on Radiation Protection & Measurements report in 2009. The report showed that Americans' average total yearly radiation exposure had increased from 3.6 millisievert to 6.2 mSv per year since the early 1980s, due to medical-related procedures. (Industrial radiation exposure, including that from nuclear power plants, is less than 0.1% of overall public radiation exposure.)

Health Physics

Health physics is also known as Physical Hazards or Physical Safety or Radiation Safety or Radiation Protection.

- Background radiation
- Radiation protection
- Dosimetry
- Health physics
- Radiological protection of patients

Clinical Audiology Physics

- Physics and acoustics for audiology: the basic physics of sound, instrumentation, and the principles of digital signal processing involved in audiological research. Topics include the physics of sound waves, room acoustics, the measurement of reverberation time; the nature of acoustic impedance; the nature of filters and amplifiers, acoustics of speech, calibration.

Laser Medicine

- Lasers and applications in medicine

Medical Optics

- Optical imaging

Neurophysics

Neurophysics (or neural physics) is also known as functional neuroimaging. It refers to imaging structure and function in the nervous system as well as:

- fMRI
- molecular imaging
- electrical impedance tomography

- diffuse optical imaging
- optical coherence tomography
- dual energy X-ray absorptiometry
- image-guided surgery

Cardiophysics

- Cardiophysics (or cardiovascular physics) is using the methods of, and theories from, physics to study cardiovascular system at different levels of its organisation, from the molecular scale to whole organisms.

- Electrocardiography

Physiological Measurement Techniques

Physiological measurements have also been used to monitor and measure various physiological parameters. Many physiological measurement techniques are non-invasive and can be used in conjunction with, or as an alternative to, other invasive methods.

- Electroencephalography
- Electromyography
- Electronystagmography
- Endoscopy
- Medical ultrasonography
- Non-ionising radiation (lasers, ultraviolet etc.)
- Near infrared spectroscopy
- Pulse oximetry
- Blood gas monitor
- Blood pressure measurement

Physics of the Human and Animal Bodies

(8.1) *Biomechanics*

- Body mass index
- Sports biomechanics

- Biofluid mechanics

(8.2) *Bioelectromagnetics* of the human and animal bodies

- Bioelectrodynamics

- Biomagnetism

- Electromagnetic radiation and health

- Electrophysiology

(8.3) *Biomaterials and artificial organs*

- Biomaterials for cell and organ therapies

- Cardiovascular biomaterials, artificial heart and cardiac assist devices

- Artificial skin, bones, joints, teeth and related biomaterials

- Histopathology

- Drug delivery and control release

Healthcare Informatics and Computational Physics

- Information and communication in medicine

- Medical informatics

- Image processing, display and visualization

- Computer-aided diagnosis

- Picture archiving and communication systems (PACS)

- Standards: DICOM, ISO, IHE

- Hospital information systems

- e-Health

- Telemedicine

- Digital operating room

- Workflow, patient-specific modeling

- Medicine on the Internet of Things

- Distant monitoring and telehomecare

Areas of Research and Academic Development

Non-clinical physicists may or may not focus on the above areas from an academic and research point of view, but their scope of specialization may also encompass lasers and ultraviolet systems (such as photodynamic therapy), fMRI and other methods for functional imaging as well as molecular imaging, electrical impedance tomography, diffuse optical imaging, optical coherence tomography, and dual energy X-ray absorptiometry.

Education and Training

In Australia and New Zealand

The Australasian College of Physical Scientists and Engineers in Medicine (ACPSEM) is the professional body that oversees the education and certification of medical physicists in Australia and New Zealand and has a mission to advance services and professional standards in medical physics and biomedical engineering.

In Europe

The presence of Medical Physicists at Expert level ('Medical Physics Experts') in healthcare in Europe is required by EC Directive 2013/59/EURATOM. The European Federation of Organizations for Medical Physics (EFOMP) has defined a detailed inventory of learning outcomes for Medical Physics Experts in terms of Knowledge, Skills and Competences (the latter in Europe means 'responsibilities'). In Europe the professional preparation for Medical Physicists consists of a first degree in Physics or equivalent (e.g., biophysics, electrical or mechanical engineering), a Masters in Medical Physics and a 2-year training Residency. For the latest EFOMP policy statement on the role and education and training requirements of the Medical Physicist and Medical Physics Expert go to. For the final version of the EC funded document 'European Guidelines on the Medical Physics Expert' download Report 174 and its annexes from.

In North America

In North America, medical physics training is offered at the master's, doctorate, post-doctorate and/or residency levels. Also, a professional doctorate has been recently introduced as an option. Several universities offer these degrees in Canada and the United States.

As of October 2013, over 70 universities in North America have medical physics graduate programs or residencies that are accredited by *The Commission on Accreditation of Medical Physics Education Programs* (CAMPEP). The majority of residencies are therapy, but diagnostic and nuclear are also on the rise in the past several years.

Professional certification is obtained from the American Board of Radiology (for all 4 areas), the American Board of Medical Physics (for MRI), the American Board of

Science in Nuclear Medicine (for Nuc Med and PET), and the Canadian College of Physicists in Medicine. As of 2012, enrollment in a CAMPEP-accredited residency or graduate program is required to start the ABR certification process. Starting in 2014, completion of a CAMPEP-accredited residency will be required to advance to part 2 of the ABR certification process.

United Kingdom

From October 2011 as part of the Modernising Scientific Careers scheme, the route to accreditation as a medical physicist in England and Wales is provided by the Scientist Training Programme (STP). This scheme is a three-year graduate scheme provided by the National School of Healthcare Science. Entrants are required to have a prior undergraduate degree (1:1 or 2:1) in an appropriate physical science.

The STP involves a part-time MSc in Medical Physics (provided by either King's College London, University of Liverpool or University of Newcastle) in addition to practical training within the National Health Service. Assessment is provided by the completion of competencies and by a final assessment similar to the OSCE undertaken by other clinical staff. Completion of the STP leads to accreditation with the Institute of Physics and Engineering in Medicine (IPEM) and registration as a Clinical Scientist.

Prior to 2011 the training route in the United Kingdom was administered in two parts, and this scheme is still used in Scotland and Northern Ireland). Part I involved limited clinical experience and a full-time MSc in medical physics. Part II involved exclusively clinical experience in which the candidate would produce a portfolio of experience which would be submitted to the Academy for Healthcare Science which (in addition to a viva) would lead to professional accreditation with IPEM.

International

There are regular regional and international educational medical physics activities. The oldest of these is the International College on Medical Physics at the International Centre for Theoretical Physics (ICTP), Trieste, Italy. This College has educated more than 1000 medical physicists from developing countries.

Legislative and Advisory Bodies

- ICRU: International Commission on Radiation Units and Measurements

- ICRP: International Commission on Radiological Protection

- NCRP: National Council on Radiation Protection & Measurements

- NRC: Nuclear Regulatory Commission

- FDA: Food and Drug Administration

- IAEA: International Atomic Energy Agency

Cell Biophysics

Cell biophysics (or cellular biophysics) is a sub-field of biophysics that focuses on physical principles underlying cell function. Sub-areas of current interest include statistical models of intracellular signaling dynamics, intracellular transport, cell mechanics (including membrane and cytoskeletal mechanics), molecular motors, biological electricity and genetic network theory. The field has benefited greatly from recent advances in live-cell molecular imaging techniques that allow spatial and temporal measurement of macromolecules and macromolecular function. Specialized imaging methods like FRET, FRAP, photoactivation and single molecule imaging have proven useful for mapping macromolecular transport, dynamic conformational changes in proteins and macromolecular interactions. Super-resolution microscopy allows imaging of cell structures below the optical resolution of light. Combining novel experimental tools with mathematical models grounded in the physical sciences has enabled significant recent breakthroughs in the field. Multiple centers across the world are advancing the research area

References

- J. M. Seddon, R. H. Templer. Polymorphism of Lipid-Water Systems, from the Handbook of Biological Physics, Vol. 1, ed. R. Lipowsky, and E. Sackmann. (c) 1995, Elsevier Science B.V. ISBN 0-444-81975-4

- "CAMPEP Accredited Professional Doctorate Programs in Medical Physics". CAMPEP. CAMPEP. Retrieved 2015-03-29.

- Hill R, Healy B, Holloway L, Kuncic Z, Thwaites D, Baldock C (2014). "Advances in kilovoltage x-ray beam dosimetry". Physics in Medicine and Biology. 59 (6). Bibcode:2014PMB....59R.183H. doi:10.1088/0031-9155/59/6/R183. PMID 24584183.

- Guibelalde E., Christofides S., Caruana C. J., Evans S. van der Putten W. (2012). Guidelines on the Medical Physics Expert' a project funded by the European Commission

Permissions

Index